BIOMIMETICS

Design and Processing of Materials

AIP Series in Polymers and Complex Materials

BOOKS IN SERIES

Statistical Physics of Macromolecules, by Alexander Yu. Grosberg and Alexei R. Khokhlov

Biomimetics: Design and Processing of Materials, edited by Mehmet Sarikaya and Ilhan A. Aksay

BIOMIMETICS

Design and Processing of Materials

Editors

Mehmet Sarikaya

University of Washington
Department of Materials Science and Engineering
Seattle, Washington

Ilhan A. Aksay

Department of Chemical Engineering
and Princeton Materials Institute
Princeton University
Princeton, New Jersey

American Institute of Physics **Woodbury, New York**

AIP Press
American Institute of Physics
500 Sunnyside Boulevard
Woodbury, NY 11797-2999

Library of Congress Cataloging-in-Publication Data
Biomimetics: design and processing of materials / edited by Mehmet Sarikaya, Ilhan A. Aksay.
p. cm. -- (AIP series in polymers and complex materials)
p. cm. -- (AIP series in polymers and complex materials)
Includes bibliographical references and index.
ISBN 1-56396-196-2
1. Biomimetics. 2. Biomimetic polymers. I. Sarikaya, Mehmet. II. Aksay, Ilhan A. III. Series.

QP517.B56B57 1995
574'.01'1--dc20 95-14071
CIP

10 9 8 7 6 5 4 3 2 1

TABLE OF CONTENTS

List of Contributors vii

Series Preface ix

Preface xi

What We Can Learn from Soft Biomaterials and Structures 1
Stephen A. Wainwright

Hierarchical Structure of Collagen Composite Systems:
Lessons from Biology 13
Eric Baer, James J. Cassidy, and Anne Hiltner

Nacre: Properties, Crystallography, Morphology, and Formation 35
Mehmet Sarikaya, Jun Liu, and Ilhan A. Aksay

Biomineralization, the Inorganic–Organic Interface, and
Crystal Engineering 91
Stephen Mann

Microstructure–Property Relations in Vertebrate
Bony Hard Tissues: Microdamage and Toughness 117
John D. Currey, Peter Zioupos, and Andy Sedman

Biomimetic Ceramics and Hard Composites 145
Paul Calvert

Microstructure of an Insect Cuticle and Applications to
Advanced Composites 163
Stephen L. Gunderson and Rebecca C. Schiavone

Structure and Function of Magnetosomes in Magnetotactic Bacteria 199
Richard B. Frankel and Dennis A. Bazylinski

Size-Quantized Particles at Artificial Membrane Interfaces 217
Janos H. Fendler

The Macromolecular Design of Spiders' Silks 237
*John Gosline, Cynthis Nichols, Paul Guerette, Audrey Cheng,
and Steve Katz*

Role of Molecular Genetics in Polymer Materials Science 263
Maurille J. Fournier, Thomas L. Mason, and David A. Tirrell

Subject Index 277

LIST OF CONTRIBUTORS

AKSAY, ILHAN A., Department of Chemical Engineering and Princeton Materials Institute, Princeton University, Princeton, New Jersey 08544-5263

BAER, ERIC, Department of Macromolecular Science, Case Western Reserve University, Cleveland, Ohio 44106

BAZYLINSKI, DENNIS A., Department of Chemistry and Chemical Engineering, Stevens Institute of Technology, Castle Point on the Hudson, Hoboken, New Jersey 07030

CALVERT, PAUL, Department of Materials Science and Engineering, Arizona Materials Laboratories, University of Arizona, Tuscon, Arizona 85712

CASSIDY, JAMES J., Department of Macromolecular Science, Case Western Reserve University, Cleveland, Ohio 44106

CHENG, AUDREY, Department of Zoology, University of British Columbia, Vancouver, British Columbia V6T 1Z4, Canada

CURREY, JOHN D., Department of Biology, University of York, York YO1 5DD, United Kingdom

FENDLER, JANOS H., Department of Chemistry, Syracuse University, Syracuse, New York 13244-4100

FOURNIER, MAURILLE J., Department of Biochemistry and Molecular Biology, Program in Molecular and Cellular Biology, Lederle Graduate Research Center, University of Massachusetts, Amherst, Massachusetts 01003

FRANKEL, RICHARD B., Department of Physics, California Polytechnic State University, San Luis Obispo, California 93407

GOSLINE, JOHN, Department of Zoology, University of British Columbia, Vancouver, British Columbia V6T 1Z4, Canada

GUERETTE, PAUL, Department of Zoology, University of British Columbia, Vancouver, British Columbia V6T 1Z4, Canada

GUNDERSON, STEPHEN L., University of Dayton Research Institute, 300 College Park, Dayton, Ohio 45469-0168

HILTNER, ANNE, Department of Macromolecular Science, Case Western Reserve University, Cleveland, Ohio 44106

KATZ, STEVE, Department of Zoology, University of British Columbia, Vancouver, British Columbia V6T 1Z4, Canada

LIU, JUN, Department of Materials Science and Engineering, University of Washington, Seattle, Washington 98195

MANN, STEPHEN, School of Chemistry, University of Bath, Bath BA2 7AY, United Kingdom

MASON, THOMAS L., Department of Biochemistry and Molecular Biology, Program in Molecular and Cellular Biology, Lederle Graduate Research Center, University of Massachusetts, Amherst, Massachusetts 01003

NICHOLS, CYNTHIS, Department of Zoology, University of British Columbia, Vancouver, British Columbia V6T 1Z4, Canada

SARIKAYA, MEHMET, Department of Materials Science and Engineering, University of Washington, Seattle, Washington 98195

SCHIAVONE, REBECCA C., University of Dayton Research Institute, 300 College Park, Dayton, Ohio 45469-0168

SEDMAN, ANDY, Department of Biology, University of York, York YO1 5DD, United Kingdom

TIRRELL, DAVID A., Department of Polymer Science and Engineering, Program in Molecular and Cellular Biology, Lederle Graduate Research Center, University of Massachusetts, Amherst, Massachusetts 01003

WAINWRIGHT, STEPHEN A., Department of Zoology, Duke University, Durham, North Carolina 27706

ZIOUPOS, PETER, Department of Biology, University of York, York YO1 5DD, United Kingdom

SERIES PREFACE

Complex materials are multi-component systems such as polymers, colloidal particles, micelles, membranes, foams, etc.; they are distinguished from simple crystalline solids and simple liquids in that they generally possess molecular or structural length scales much greater than atomic. They often exhibit unusual properties or combinations of properties, such as light weight combined with strength, toughness combined with rigidity, fluidity combined with solid-like structure, or the ability to dissolve in both oil and water. Complex materials turn up as surfactants in soaps and facial creams, as lubricants on magnetic disks, as adhesives in medical supplies, as encapsulants in pharmaceuticals, etc. Polymers are ubiquitous as fibers, films, processing aids, rheology modifiers, and as structural or packaging elements.

Some recent additions to the list of complex fluids include liquid crystalline polymers, sheet-like polymers, polymers with non-linear optical properties, bicontinuous block copolymers and interpenetrating networks, advanced zeolites and molecular sieves made from mixtures of liquid crystals and silicates, and new and exotic chiral liquid crystalline phases.

The study of complex materials is highly interdisciplinary, and new findings are published in a bewildering range of journals and periodicals and in many scientific and engineering societies, including ones devoted to physics, chemistry, ceramics, plastics, material science, chemical engineering, and mechanical engineering. Experimental techniques used to study polymers and complex fluids include scattering of neutrons, x-rays, and light, surface-force measurements, cryo-microscopy, atomic force microscopy, magnetic resonance imaging, and many others. The aim of this series is to bring together the work of experts from different disciplines who are contributing to the growing area of polymers and complex materials.

It is anticipated that these volumes will help those who concentrate on one type of complex material to find insight from systematic expositions of other materials with analogous microstructural complexities, and to keep up with experimental and theoretical advances occurring in disciplines other than their own. We believe that the AIP Series on Polymers and Complex Materials can help spread the knowledge of recent advances, and that experimentalists and theoreticians from many different disciplines will find much to learn from the upcoming volumes.

Ronald Larson
AT&T Bell Laboratories

Philip A. Pincus
University of California, Santa Barbara

PREFACE

Biomimetics is the study of biological structures, their functions, and their synthetic pathways in order to stimulate new ideas and to develop these ideas into synthetic systems similar to those found in biological systems. As part of this endeavor, learning lessons from biology has spurred interest across many disciplines. In recent years translating these lessons into new materials development has constituted a rapidly growing field of interest within the materials community. In many ways, biomimetics symbolizes the true nature of materials science as a multidisciplinary field, joining diverse disciplines ranging from biology and biochemistry to physics and mathematics. The chapters of this book were prepared by some of the leading researchers from these disciplines in order to produce a comprehensive review of this new field and provide a resource for physical scientists and biologists alike in their current and future research endeavors.

The first chapter by Stephen A. Wainwright introduces the philosophy of biomimetics with examples from biology in terms of both structures and functions. Eric Baer, James J. Cassidy, and Anne Hiltner in the second chapter introduce the principles of hierarchy in soft connective tissues. Mehmet Sarikaya, Jun Liu, and Ilhan A. Aksay in the third chapter give a review of relatively simple structures of laminated hard tissues, with emphasis on the nacre of mollusk shells. In the fourth chapter, Stephen Mann emphasizes the concepts and strategies of biomineralization. In the fifth chapter, John D. Currey, Peter Zioupos, and Andy Sedman review properties of a complex biological hard tissue, bone. In the sixth chapter, Paul Calvert introduces some processing routes for ceramic-based materials. In the seventh chapter, Stephen L. Gunderson and Rebecca C. Schiavone review the structures of insect cuticles with an emphasis on design for engineering applications. In the eighth chapter, Richard B. Frankel and Dennis A. Bazylinski discuss the current understanding of ultrafine inorganic particle formation in biological systems (bacteria); and in the ninth chapter Janos H. Fendler follows this up on some of the membrane-mimetic pathways for the formation of semiconductive and optical particles. John Gosline and his colleagues in the tenth chapter discuss the structure and properties of various silks produced by spiders. In the last chapter, the book is concluded by Maurille J. Fournier, Thomas L. Mason, and David A. Tirrell who discuss strategies for genetically engineered polymers.

We express our appreciation to Frederick Hedberg and George Haritos of the U. S. Air Force Office of Scientific Research for their support of biomimetics research. A workshop organized by them led to the compilation of the chapters that appear in this book. Without their encouragement and the able assistance of Melba Wallace, Daniel Dabbs, and Arla Dittrick this book could not have been completed.

Mehmet Sarikaya
Ilhan A. Aksay

WHAT WE CAN LEARN FROM SOFT BIOMATERIALS AND STRUCTURES

Stephen A. Wainwright

Department of Zoology, Duke University
Durham, North Carolina 27706, USA

Future design of engineering composite materials may be based on the structures and functions of biological soft and hard tissues. These biological materials include soft tissues such as mucus, cartilage, tendon, and skin; and hard tissues such as skeletal units, teeth, mollusk shells, and scales. The uniqueness of biological materials comes from several facts including their complex and intricate, ordered structures and their multifunctionality. Biomimetics offers the investigation of the structures and functions of these biological materials and allows possible future design and synthesis of engineering composites based on the principles obtained from the biological materials. This article describes several biological tissues and discusses their unique physical properties as a guide to design criteria for new engineering materials.

1.0 THE ROLE AND PROMISE OF BIOMIMETICS

Biomimetics will engulf molecular biology and replace it as the most challenging and important biological science of the 21st Century. This will be the logical extension of the directions and achievements in the engineering design of composite materials and robots in the late 20th Century and the requirements of fabricational technology of the future. We are now learning to produce more complex materials and machines inspired by the ways that nature does: this will keep us on the ascending part of our cultural learning curve for the next hundred years. Molecular biology will contribute to this goal, but the goal will only be achieved if we put renewed vigor into basic research on the mechanisms of function of cells, tissues, organs, organ systems, and the organisms of which they are parts. Even more important will be biomimetics-oriented research on the processes of fabrication (= development) and regeneration of materials and systems in plants and animals.

The most remarkable and technologically attractive differences between natural and synthetic materials are the great hierarchical complexity and multifunctionality of natural materials. A few selected examples of structural biomaterials and systems will illustrate this. These examples are bound also to stimulate our thoughts on what they can teach us about materials of the future.

2.0 IONIC POLYMERIC MATERIALS

There is a continuum of soft, polymeric materials produced by animals and plants, from the mucus of seaweeds and slugs, including all the soft collagenous connective tissues throughout our body, to the cartilage of our long bones, whose properties can be altered by changes in their ionic environment. Most intriguing are the sea urchins and their relatives, because they have physiological control over this environment and can cause their ligaments to be stretchy or stiff according to their immediate needs.

2.1 MUCUS: SOLID-FLUID TRANSFORMATION

What is involved in highly hydrated materials like mucus is a mixture of two parts: one part is a high molecular weight glycoprotein shaped like a bottle brush (Figure 1). Each "bristle" of the brush is a side group that has a strong negative ionic charge. The long protein molecule forming the core and stem of the "brush" can be tightly covalently linked with others of its kind into enormous clusters. The second part is the aqueous ionic solution surrounding the molecular clusters. The negatively charged side groups repel each other with force and thus imbibe water, inflating the molecule. So a cluster of these molecules holds a volume of water much larger than the molecules themselves. Anions in the water can reduce the imbibing force by competing with water ions for the negative charges on the glycoprotein.

The clusters may interact with each other by forming weak hydrophilic bonds. So a thin layer of this material between your thumb and finger may act as a glue. If you overpower the glue and shear the layer, it suddenly loses its stiffness and becomes a lubricating fluid. Stop moving your fingers and it seizes up again into a glue. You may repeat this as often as you like. Slugs use mucus in this way on the sole of their foot to allow them to stick to the ground at the same time that they slide over it.[1] Repeated

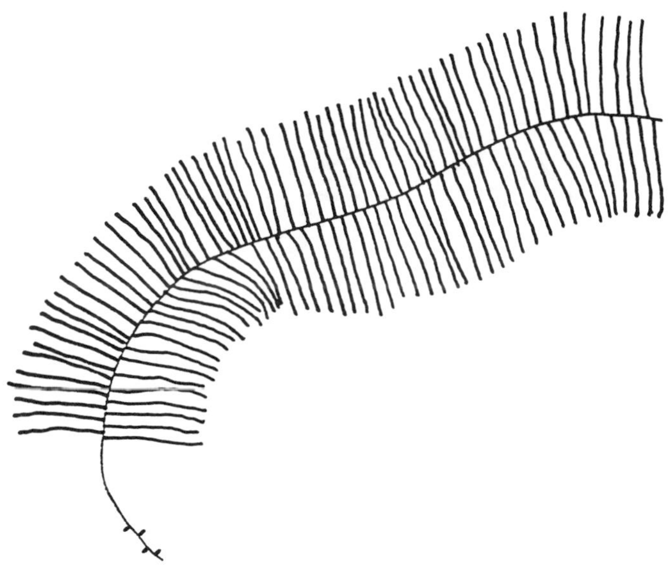

Figure 1. A mucus molecule has a long central protein core and an array of negatively charged polysaccharide chains attached to it. This figure is 2-dimensional: the molecule is 3-d and more like a bottle brush than a feather.

transformation from fluid to solid and back again is a cute trick we could make use of more often.

2.2 MESOGLEA OF SEA ANEMONES: ANISOTROPIC MATERIAL

Mesoglea is the soft, collagenous connective tissue that forms the skeletal integrity of squishy animals like sea anemones and jellyfish. Mesoglea is a mucus-like material surrounding and linked to a set of collagen fibers, all immersed in an ionic fluid that is very like sea water.[2] When the collagen has parallel orientation wrapped circumferentially around the anemone's cylindrical body, the high tensile stiffness of collagen-slime complex renders the tissue very much stretchier in the tallness direction of the anemone than in the girth direction (Figure 2). This prevents explosions of the body produced by hydrostatic pressures created by muscles in the body wall. (In pressurized cylinders, girth stresses are twice longitudinal stresses simply because of geometry.)[3]

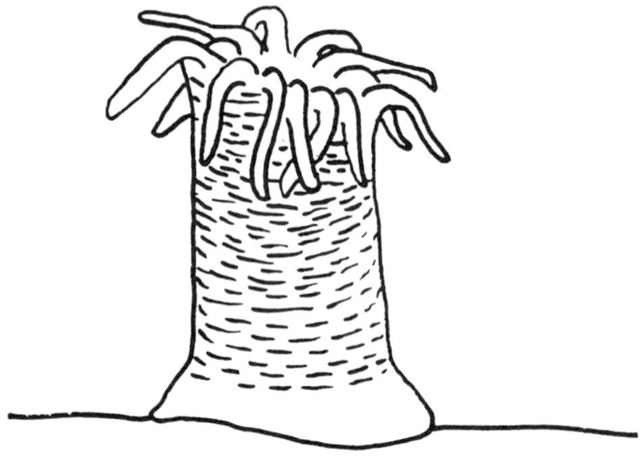

Figure 2. The inner layer of connective tissue of sea anemones contains a circumferentially oriented array of collagen fibers that allows great changes in height but limits girth.

2.3 ARTICULAR CARTILAGE: HYDRAULIC SHOCK ABSORBER

In the cartilage at the ends of one's long bones, this system is much more complex.[4,5] When one stands up, the weight held up by the thigh bones bears at a few small places on the top of the shin bones. As the weight shifts forward when one takes a step, the cartilage gets squeezed between the thigh and shin bones. Within the cartilage, compressive force tends to squeeze the ionic fluid away from the pressure points (Figure 3). The big proteoglycan molecules are swept along too, but they are caught by the tight network of collagen molecules. So the proteoglycan hung up on the collagen network creates a drag force that retards the flow of fluid. When the force is removed, the big molecules stay put and imbibe the watery fluid back into its original position. This means that one's weight is supported on a watery pad and that any impact, such as your jumping, will be damped by the viscous flow of the ionic fluid through the molecular spongework. This is a remarkable safety feature: imagine how long we would last if brittle bone met brittle bone at every step, without padding or damping!

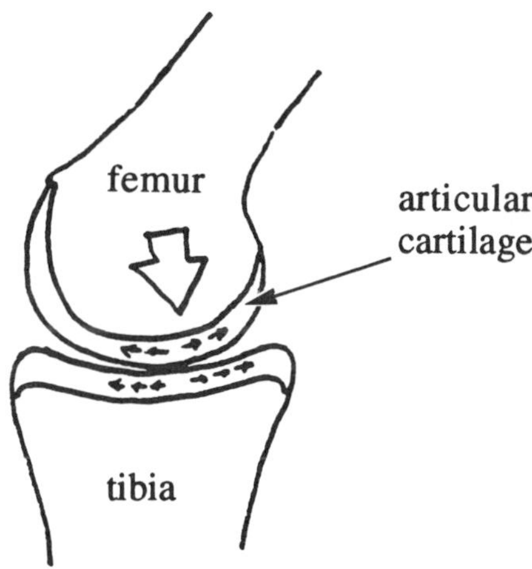

Figure 3. Where our leg bones meet, articular cartilage protects them. At the point of contact, compressive force (hollow arrow) causes fluid in the cartilage to flow away from the force. When the force is released, the fluid is imbibed by charged molecules to return.

2.4 CATCH CONNECTIVE TISSUES IN SEA URCHINS: CONTROLLABLE PROPERTIES

A sea urchin can wave its stiff, sharp, calcitic spines at us by using muscles around the joint at the base of each spine (Figure 4). There are collagenous ligaments, lying in parallel with the muscle, that connect each spine to the rigid shell of the urchin. While the spine is being waved around, the ligaments must stretch. But when we touch the urchin, the ligaments lose their extensibility and become rigid in tension so that we cannot bend the spine's joint in any direction: the spine breaks instead. After we tire of breaking spines and go away, the urchin will "loosen" these ligaments again and wave the spines at the next intruder.

The urchin's spine ligament consists of parallel collagen fibers interpenetrated by extensions of neurosecretory cells of a nervous ganglion that sits on the ligament. The fingerlike neuronal extensions (axons) contain large, plump vesicles that flow along the axon and may be secreted out into the ligament. We know from experiments on iso-

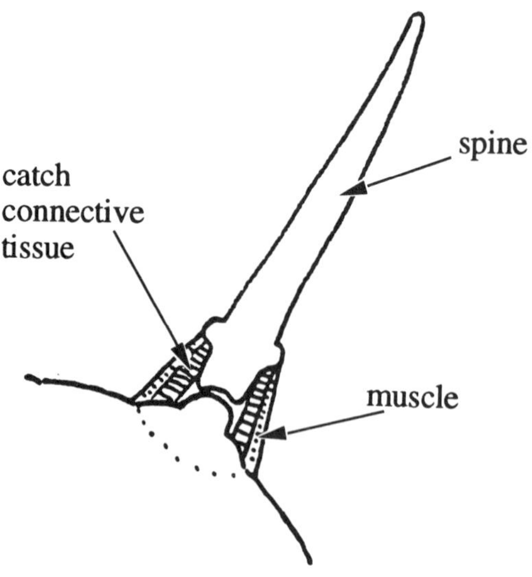

Figure 4. The spine of a sea urchin, shown here in longitudinal section, is held to the skeleton by a ring of muscle and a ring of catch connective tissue.

lated ligament preparations that the stiffness of the ligament varies with the concentration of calcium ions. We are still learning about the control mechanisms.[6,7] With this ability to control the stiffness of its structural materials, the urchin can use the sharp spines as mobile legs for walking and it can wave the spines to ward off predators. It can also wedge itself tightly into a rocky crevice where predators cannot dislodge it. If we could take a lesson from the urchin about controlling the mechanical properties of connective tissues, and add it to our knowledge of cartilage and other connective tissues, we should be able to invent ways to restore, pharmacologically, connective tissue properties within our own diseased bodies.

2.5 MULTIFUNCTIONAL WHALE BLUBBER

Whale blubber is perhaps the most multifunctional material we know. Its fat content makes it a floatation device (whales breathe air), an excellent insulation against the infinite heat sink of the ocean (whales are warm blooded), and it is the best possible food

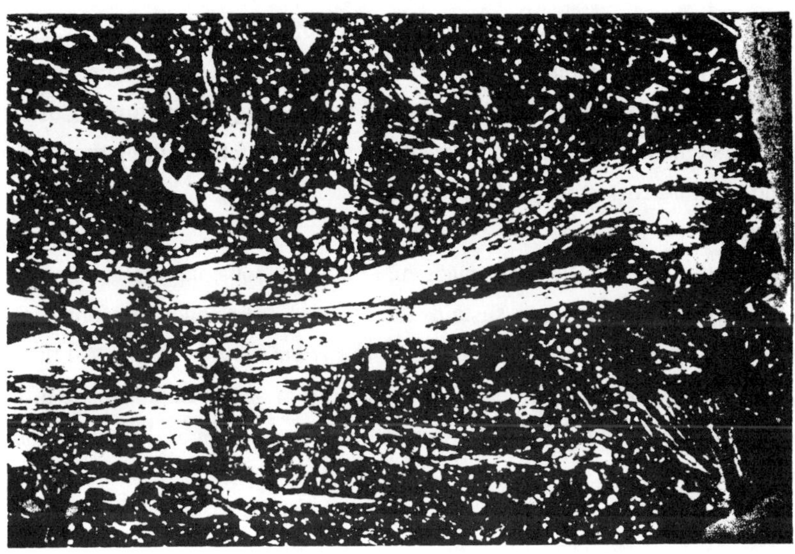

Figure 5. Micrograph of a section through the blubber of a whale. The fibers are collagen (~ 0.1 mm diameter) and fat is contained in the cells shown.

reserve for use during 6,000 km long, nonfeeding migrations (fat yields 3 times more metabolic energy per gram than protein or polysaccharide). Blubber is also a very bouncy rubberlike material. Our best estimate now is that acceleration caused by the elastic recoil of blubber that is compressed and stretched with each tail stroke may save up to 20% of the cost of locomotion during extended periods of continuous swimming.[8]

We have only recently learned that while the blubber of whales and dolphins contains much fat, it also contains up to 50% by volume of collagen. The collagen is in large fibers, visible to the naked eye, that are oriented in crossed helical patterns winding around the animal's body and another set oriented radially, inside to outside, through the thickness of the blubber layer (Figure 5).

This is again a frontier. We do not yet know the function of the radial collagen fibers and we have no information on how the fiber array changes with motion or how the fibers interact with the fat. If we knew these things, we can be sure that we would find a use for a material with these properties somewhere in our industrial or medical world.

The amount (volume percent) and the winding angles of collagen in blubber vary from place to place in the animal causing concomitant variation in mechanical properties. For example, an imaginary belt of blubber around the flexing, locomotor part of the body varies in its tensile modulus and extensibility: it is stretchiest along the top and bottom of the belt and stiffest along the sides.[8,9]

This means that, unlike a classic beam made of a homogeneous material where peripheral material in the bending plane sustains the greatest strains and therefore stresses, the bent hollow cylinder of blubber is being *evenly strained and stressed.* All the material is being used equally, not just the top and bottom parts. Beams made of a graded material that distribute bending stresses evenly to all parts of the beam would be very useful.

2.6 COMPLEX POLYMERIC MOTOR SYSTEMS: THE ELEPHANT'S TRUNK

Typical robotic arms and manipulators have several stiff segments separated by flexible joints. For every joint there must be a motor and for every motor a control "box." For a multi-jointed finger, there must be a control box for all the control boxes. Would not it be simpler and cheaper if the finger was a single hollow tube, like the finger of a glove, that bent when pressurized? Such manipulative opposing fingers and their hydrostatic supporting arm were designed and built by Jim Wilson and his students in consultation with my students and me. Our zoological role was to help them understand how the elephant's trunk works.[10]

Elephant trunks, tentacles of squids and octopi, and tongues of people and chameleons are all muscular hydrostats.[11] They have no rigid cartilage or bony materials, nor do they have large fluid-filled cavities. They consist of soft, collagenous connective tissues and a considerable bulk of the muscle that drives them. Muscle is a tissue made up of long, thin cells. Each cell contains the force-producing contractile protein conglomerate, actomyosin, as well as nuclei, mitochondria, and other inclusions all suspended in an aqueous fluid.

In the elephant's trunk, the muscles that cause it to curl are oriented longitudinally. There are also radial and transverse muscles throughout the trunk that stabilize curling and control the change in length. But the muscles are small and there are literally thousands of them. Exactly how the nervous system controls all the muscles in the trunk has not even been studied yet. So Wilson did not try to make his manipulator exactly like the elephant's.[10] He got the general idea for a hydrostatic manipulator from reading about worms, squids, and elephants. He and his students then designed a cylindrical arm with one control unit with a second control unit for the cylindrical fingers.

Design of the fingers led to the use of a bellows structure for the outer side of the bending finger. Wilson found no general treatment of the mechanics of bellows in the library, so he wrote one along the way.[10] He also calculated that the best viscoelastic properties of the polymeric material for the finger would be different from those of plastics on the market. He gave his design requirements to a plastics fabricator and caused a new material to be designed and made for this new use. This is an example of a phenomenon that will surely be repeated often by many industries in the years to come. It is an important aspect of the promise of biomimetics.

Finally, in the design of the arm, he followed the form of bending actually found in the elephant's trunk which he learned about from movies we took of highly intelligent and co-operative animals lifting weights in the National Zoo in Washington, D.C. Compared to a simple, jointed, multimotored robotic manipulator, Wilson's gadget and its control system weigh one tenth the jointed one, it performs a simple action (picking up a block from one shelf and placing it on another) in one tenth the time taken by the jointed one, and the cost of production was a hundredth that for the jointed one. Comparing it to the elephant's trunk, Wilson's manipulator is actuated by forced pressure changes, not by forceful shortening of a polymer as in the elephant. So he took a little instruction and a considerable amount of inspiration from nature.

3.0 THOUGHTS AND CONCLUSIONS

Among the soft tissues in plants and animals there is a great diversity of materials varying in complexity from slightly complex (slime), to more complex (8 levels in the rat tail tendon),[12] to so complex that we have not yet even counted the levels of structural integration (blubber). These materials have equally diverse mechanical, ionic, magnetic, and heat and light conducting properties. Many biomaterials have multiple uses involving different physical modes, and all biomaterials are produced by small, wet, polymeric fabricational units (cells) at room temperature and pressure from simple molecular building blocks taken from other plants and animals and trace elements that are available in the earth, air, and seas of our Earth.

We humans also fabricate materials from components we find in nature. However, our fabricated building materials are structurally simple and are most often used in single purpose designs. The production of a useful polymer from petroleum we pump out of the ground is expensive and involves high heat and pressure.

It seems reasonable that we can make great use of a more profound knowledge than we have at present of biomaterials in the design and production of synthetic materials in the 21st Century. We must learn much more about the relationships between structure and physical properties of natural, hierarchically complex, polymeric materials (not to mention the things we can learn about ceramics from shells and bones.[13-15] And then we must learn how natural materials are produced, how they become structurally organized, how the components are linked, and how structure at each level in the hierarchy contributes to the properties of the material in the performance of the organ of which it is a part.

Perhaps more important and far reaching are the inspirations we can get from the study of natural materials. We can decide to invent structures or fabricational processes that perform like the ones we see in nature. And as we invent more complex instruments for space and deep oceanic exploration and for medically exploring and manipulating the normal and ailing human body, we will see the need for more complex materials to put in them. We can expect to invent the new materials, not necessarily identical to, but analogous to those in plants and animals.

These thoughts and design processes are already happening, and the number and magnitude of such synthetic hierarchical materials is growing every year. Among the tools we will be using more and more are the control processes for polymer fabrication that we are learning from molecular biology. Biomimetics is an ideal application of this great motherlode of information.

4.0 REFERENCES

1. M. Denny and J. M. Gosline, "The Physical Properties of the Pedal Mucus of the Terrestrial Slug," *J. Exp. Biol.*, **88**, 375-393 (1980).

2. J. M. Gosline, "Structure, Composition and Viscoelastic Properties of Mesogl," *J. Exp. Biol.*, **55**, 763-795 (1971).

3. *Mechanical Design in Organisms*, S. A. Wainwright, W. D. Biggs, J. D. Currey, and J. M. Gosline (Princeton University Press, Princeton, NJ, 1982).

4. *Biomechanics of Diarthrodial Joints*, Vols. I and II, V. C. Mow, A. Ratliffe, and S. L.-Y. Woo (eds.) (Springer-Verlag, New York, 1990).

5. V. C. Mow, W. M. Lai, and J. S. Hou, "Triphasic Theory for the Swelling Properties of Hydrated Charged Soft Biological Tissues," *Appl. Mech. Rev.*, **43** [2] 134-141 (1989).

6. T. Motokawa, "Connective Tissue Catch in Echinoderms," *Biol. Rev.*, **59**, 255-270 (1984).

7. J. A. Trotter and T. J. Koob, "Collagen and Proteoglycan in a Sea Urchin Ligament with Mutable Mechanical Properties," *Cell Tiss. Res.*, **258**, 527-539 (1989).

8. L. S. Orton, "Fin Whale Mechanics: Blubber as a Structural Material," Ph.D. Dissertation, Duke University (Durham, N.C., 1988).

9. L. S. Orton and P. F. Brodie, "Whale Blubber Elasticity," *Am. Zoologist*, **25**, 13A (1986).

10. J. F. Wilson, U. Mahajan, S. A. Wainwright, and L. J. Croner, "A Continuum Model of Elephant Trunks," *J. Biomech. Engin.*, **113**, 79-84 (1991).

11. W. M. Kier, and K. K. Smith, "Tongues, Tentacles and Trunks: Biomechanics of Movement in Muscular Hydrostats," *Zool. J. Linn. Soc.*, **83,** 307-324 (1985).

12. E. Baer, J. J. Cassidy, and A. Hiltner, "Hierarchical Structure of Collagen Composite Systems: Lessons from Biology," this book, pp. 13-34.

13. J. D. Currey, P. Zioupos, and A. Sedman, "Microstructure-Property Relations in Vertebrate Bony Hard Tissues: Microdamage and Toughness," this book, pp. 113-141.

14. M. Sarikaya, J. Liu, and I. A. Aksay "Nacre: Properties, Crystallography, Morphology, and Formation," this book, pp. 35-90.

15. S. Mann, "Biomineralization, the Inorganic-Organic Interface and Crystal Engineering," this book, pp. 91-116.

HIERARCHICAL STRUCTURE OF COLLAGEN COMPOSITE SYSTEMS: LESSONS FROM BIOLOGY*

Eric Baer, James J. Cassidy, and Anne Hiltner

Department of Macromolecular Science
Case Western Reserve University
Cleveland, Ohio 44106, USA

Hierarchical structures in biocomposite systems such as in collagenous connective tissue have many scales or levels, have highly specific interactions between these levels, and have the architecture to accommodate a complex spectrum of property requirements. As examples, the hierarchical structure-property relationships are described in three soft connective tissues: tendon, intestine, and invertebral disc. In all instances, we observed numerous levels of organization with highly specific interconnectivity and with unique architectures that are designed to give the required spectrum of properties for each oriented composite system. From these lessons in biology, the laws of complex composite systems for functional macromolecular assemblies are considered.

1.0 BACKGROUND

Soft connective tissues, designed to serve specific functions in the body of human and animals, are among the most advanced structural composite materials known made of macromolecular building blocks. By utilizing the same basic macromolecular design and only by varying the hierarchical structure, a wide range of tissues possessing very different properties are synthesized by the organism. This article reviews our recent work on the hierarchical structure of soft connective tissues from the animal kingdom in hopes that these "lessons from biology" may provide polymer and materials scientists with new ideas for the design of high performance composite polymeric materials. Only with an appreciation and understanding of the unique structure-property relationships in such biosystems can this be achieved.

---

* Plenary Lecture by Eric Baer; 33rd IUPAC - International Symposium of Macromolecules; MACRO-90; Montreal, Canada, July 8-13, 1990.

All soft connective tissues have remarkably similar chemistry at the macromolecular and fibrillar levels of structure. This similitude extends through the collagen fibril which is the basic building block of all soft connective tissues. Differentiation in the hierarchical structure takes place when these fibrils are arranged in a particular architecture thus constructing a particular tissue for a unique function.[1] The hierarchical structure of a tissue reflects and depends upon the different stress states in which the tissue is required to function. Other requirements are also accommodated or coupled simultaneously such as the transport of water and the diffusion of the products of digestion. In each of these cases, the upper levels of the hierarchical structure or architecture are organized with specific mechanical and transport requirements in mind. Furthermore, the many discrete layers of the hierarchical structure must interact in such a way as to make the performance of these tasks possible.[2]

The basic structural fiber in all connective tissues is collagen. It is often the most abundant protein in animals and is widely distributed in the structural elements of the body. The common elements in the structure of soft connective tissues begin at the molecular level with similarities in the amino acid sequences of this class of proteins. This similitude lays the groundwork for the development of a variety of tissues that share common chemical and physical properties. Collagens contain the amino acid hydroxyproline, which is not commonly found in other proteins. Along with proline and glycine, these three amino acids account for more than fifty percent of the total amino acid content in all collagen types. Addition of other amino acids or variations in the ratio of these amino acids differentiate between the different collagen types. Amino acids play a major role in determining the three dimensional conformation of the precursor to collagen, the tropocollagen molecule. This molecule is a coiled-coil of three helical polypeptides 290 nm in length (Figure 1).[3] Five of these molecules align longitudinally with an overlap of approximately one quarter the molecular length to form a microfibril of diameter 3.6 nm. This so-called quarter stagger combined with the gap between successive macromolecules is responsible for the characteristic 64 nm banding pattern observed in the electron microscope and by X-ray diffraction. The microfibrils are then assembled into collagen fibrils that may vary in thickness from 35 to 500 nm (Table-I). These basic building blocks are combined, oriented, and laid up to form higher ordered structures with a particular morphology to suit the requirements of a tissue.

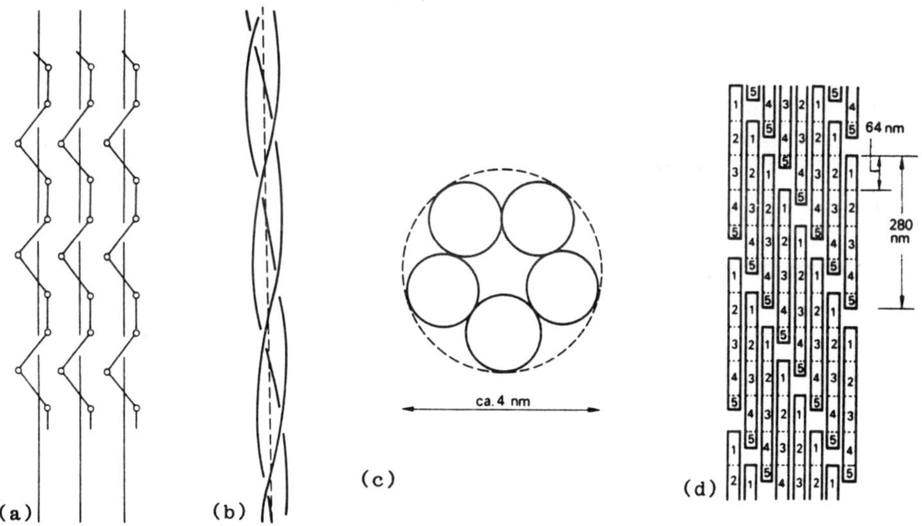

Figure 1. Building blocks of the collagen fibril. The association of amino acid molecules into the collagen fibril is illustrated schematically in these diagrams. (a) Three polypeptide chains, each coiled separately about a minor axis, form a coiled-coil (b) about a single central axis. This triple helical molecule is tropocollagen. (c) Five tropocollagen molecules aggregate to form a microfibril with diameter approximately 4 nm. (d) In a longitudinal section, the quarter staggered arrangement of the tropocollagen helices along the length of the microfibril is shown. This quarter stagger and the gap between the ends of the molecules gives rise to the 64 nm banding pattern observed in collagen fibrils (Ref. 3).

The collagen fibrils are surrounded by an extracellular matrix that maintains the integrity and architecture of the collagen. The primary component of this matrix is high molecular weight hyaluronic acid with a highly branched aggregate of proteoglycans. Proteoglycans consist of a core protein with numerous pendant mucopolysaccharide molecules. These mucopolysaccharides include chondroitin sulfates 4 and 6, and keratan sulfate. The amounts of these molecules and their ratios vary between connective tissues, with location within tissues, and with age. It is the ability of the proteoglycans to imbibe water which swells the matrix and supports the collagen fibrils. The mecha-

Table-I. Collagen Fiber Size and Distribution in Various Tissues

	Diameter (nm)	Distribution
Rat Tail Tendon		
Fetal	30	Normal (Broad)
Adult	450	Broad (Bimodal)
Human Periodontal Ligament		
Fetal	70	Normal (Narrow)
Adult	40	Normal (Narrow)
Human Intervertebral Disc		
Fetal	31	Normal
Adult	40, 100-150	Bimodal
Human Heart Valve Leaflet		
Adult	30-50	Normal
Rat Intestine (2 years)	400	Normal (Narrow)
Rabbit Cornea	20-25	Normal (Narrow)

nical properties of the matrix are regulated by its water content which, in turn, affects the properties of the composite tissue as a whole.

The overwhelming consideration in the arrangement of collagen fibrils to form connective tissues is its function in the body. This will be illustrated by three tissue types studied in our laboratories at Case Western Reserve University.

2.0 EXAMPLES OF COLLAGEN COMPOSITE SYSTEMS

2.1 TENDON

Tendons connect muscle to bone around a joint thereby converting muscle contraction into joint motion. As such, the tendon is subjected almost exclusively to uniaxial tensile stresses oriented along its length. This situation requires that the tendon be elastomeric yet sufficiently stiff to efficiently transmit the force generated by the muscle. At the same time, it must be capable of absorbing large amounts of energy without fracturing. For example, absorbing the force generated about the knee joint in a fall. It ac-

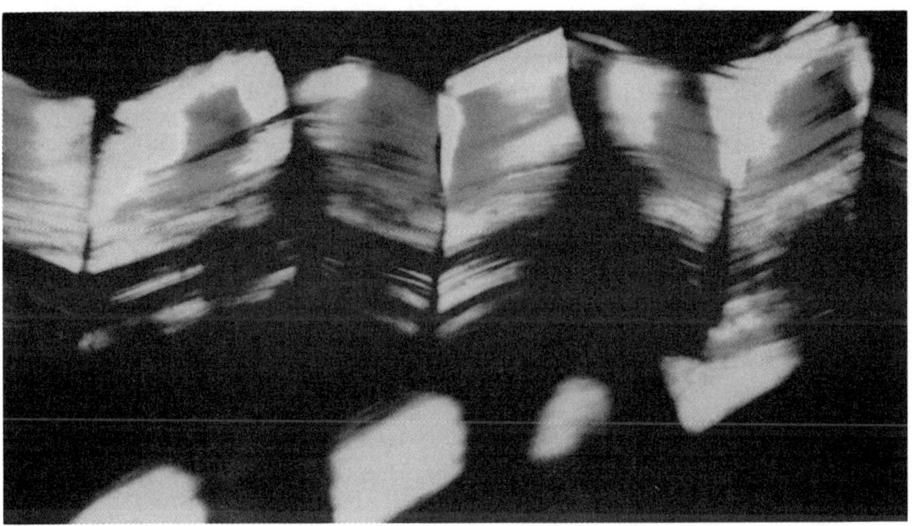

Figure 2. The morphology of the collagen fibril in tendon. Viewed between crossed polarizers in the optical microscope, the collagen fibrils that make up the tendon have a wavy appearance. Upon further examination this waveform is characterized as a planar zigzag. Adjacent fibrils are all crimped in register. It is this wavy conformation of the fibrils on the microscopic level of the structure that imparts a high degree of elasticity to the tendon, enabling it to be stretched repeatedly longitudinally without damaging the underlying structure on the nano- and molecular levels (Ref. 1). *(See also **Color Plate 1**.)*

complishes this through a unique hierarchical structure in which all levels of organization from the molecular through the macroscopic are oriented to maximize the reversible and irreversible tensile properties in the longitudinal direction without fracture.

In the tendon, collagen fibrils are organized into larger fibers. These fibrils are arrayed parallel to one another and oriented longitudinally between the muscle and bone. When the fibers are observed in between crossed polarizers in the optical microscope, they have an undulating appearance (Figure 2).[1] Further examination reveals the waveform to be a planar zigzag or crimp rather than a helix.[4] That is, the macroscopic structure does not reflect the helical conformation of the collagen macromolecules. When the tendon is stretched along its length, this crimp waveform is gradually straightened (Figure 3).[4] It is the magnitude of this waveform that determines the reversible elastic

Table-II. Crimp Parameters for Various Tissues (Ref. 5).

Source and Age	Period (μm)	Angle (deg)
Tendons:		
Rat Tail (14 months)	200	12
Human Diaphragm (51 years)	120	12
Kangaroo Tail (11.7 years)	150	8-9
Human Achilles (46 years)	40-100	6-8
Other Tissues:		
Human Periodontal Ligament (8 years)	32	25
Human Intervertebral Disc (31 & 36 years)	12-16	20-42*
Heart Valve Leaflet	20	28-30
Rat Intestine (3 month)	20	30-56
Rabbit Cornea (adult)	14	(sine wave)

* The disc is a gradient structure and crimp parameters change with radial distance through the annulus fibrosus.

properties of the tendon. As the tendon is pulled further, all the crimp in the collagen fibers is eventually pulled out. The waviness observed in collagen fibers in the rat tail tendon is also found in tendons from other species and in other connective tissue types (Table-II).[5] This generality across species and tissue lines indicates the ubiquitousness of this crimp morphology and its importance in determining the mechanical response of all soft connective tissues. As the upper levels of structural architecture are varied to meet the mechanical and other environmental requirements of a particular tissue, so are the parameters of the crimp waveform altered to adjust the mechanical response of that tissue.

Further examination of the structure of tendon for structural hierarchy reveals both the size distribution and the composite nature of the collagen fibrils. By cutting the tendon transversely, one observes that in young animals, the fibrils are of small diameter and they are all essentially the same size (Figure 4).[6] As the animal matures and ages, the fibrils get larger and the range of fibril sizes broadens considerably (Table-I). Interspersed between the fibrils is the matrix material discussed earlier. A longitudinal section through a tendon that has been extended to remove the crimp shows the 64 nm banding pattern that is characteristic of collagen and results from the staggered

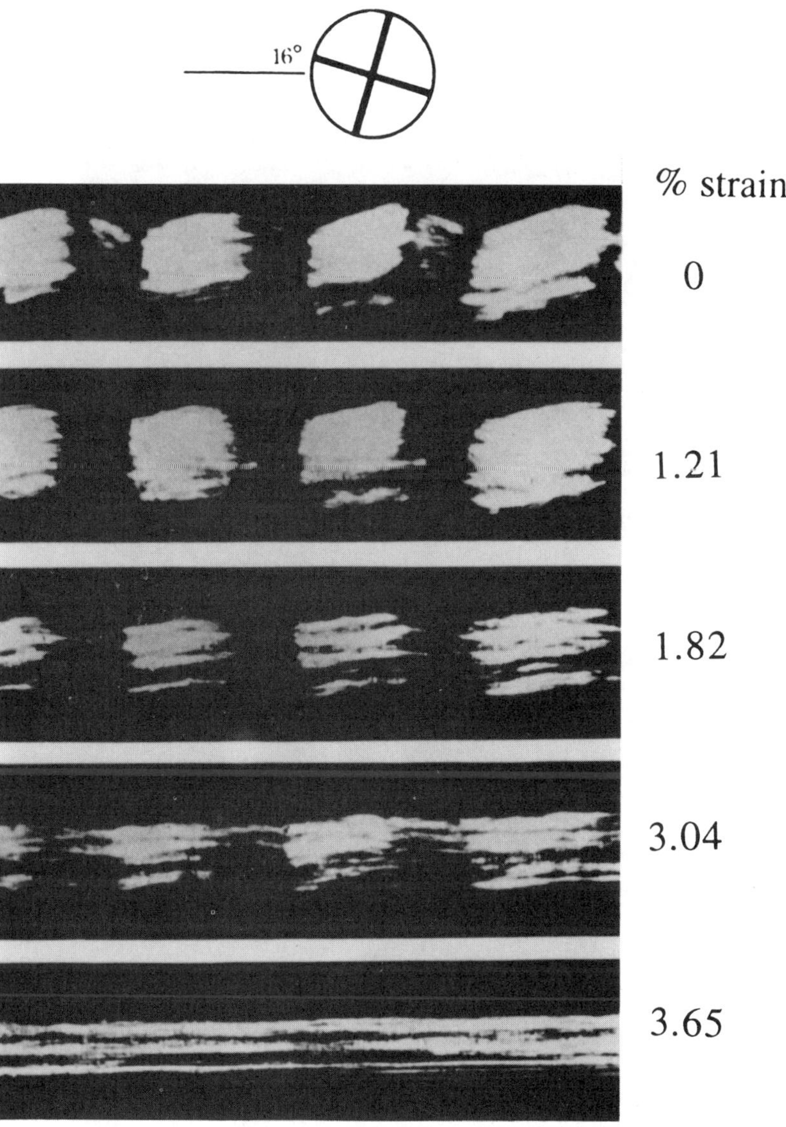

Figure 3. Crimp straightening in the tendon. The performance of the tendon in the toe region of the stress-strain curve is directly attributable to the wavy, conformation of the collagen fibrils. In this series of photomicrographs, the progressive straightening of the planar zigzag with increasing tensile strain is shown. The physiological loading range of the tendon is in this regime, so most loads are borne by this crimp waveform (Ref. 4).

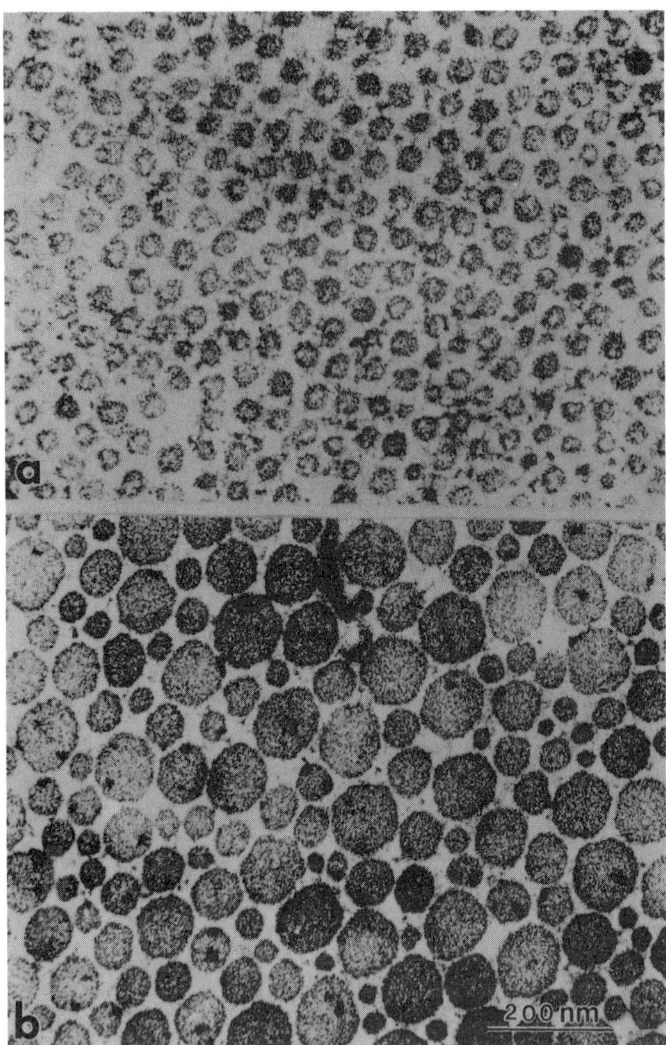

Figure 4. Electron microscopy of collagen fibers in tendon. Transverse sections of undeformed rat tail tendon show the change in fibril size and distribution with age. (a) In the newborn rat, the fibrils are of uniform size, 30 nm in diameter. (b) At 30 months, there is a distribution of fibril sizes up to 350 nm in diameter in this section (Ref. 6).

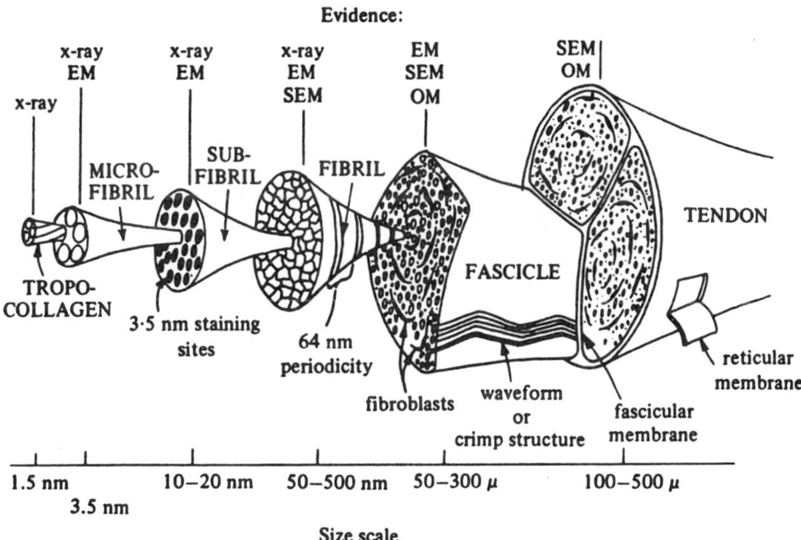

Figure 5. Hierarchical structure of the tendon. The hierarchical organization of connective tissues is illustrated in the tendon. Beginning at the molecular level with tropocollagen, progressively larger and more complex structures are built up on the nano- and microscopic scales. At the most fundamental level is the tropocollagen helix. These molecules aggregate to form microfibrils which, in turn, are packed into a lattice structure forming a subfibril. The subfibrils are then joined to form fibrils in which the characteristic 64 nm banding pattern is evident. It is these basic building blocks that, in the tendon, form a unit called a fascicle. At the fascicular level the wavy nature of the collagen fibrils is evident. Two or three fascicles together form the structure referred to as a tendon. It is this multilevel organization that imparts toughness to the tendon. If the tendon is subjected to excessive stresses, individual elements at different levels of the hierarchical structure can fail independently. In this way, the elements absorb energy and protect the tendon as a whole from catastrophic failure (Ref. 7).

arrangement of the microfibrils. It is apparent that these soft connective tissues are a hierarchical structure constructed of many collagen fibrils that are themselves composite materials (Figure 5).[7]

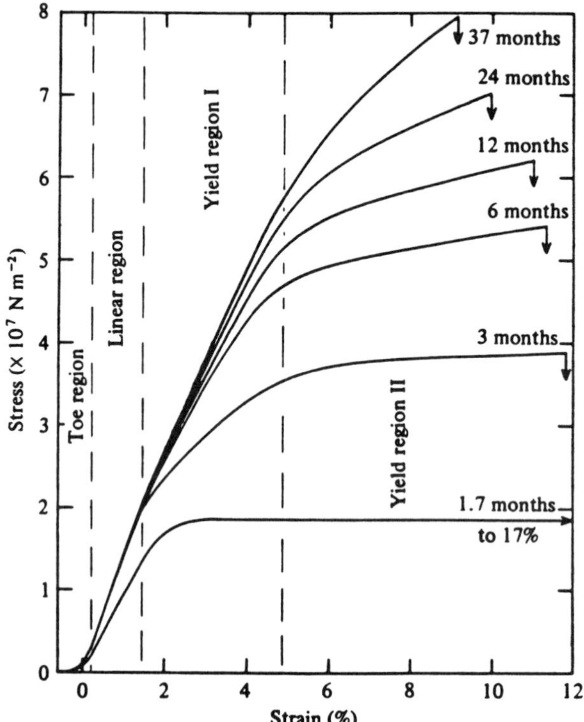

Figure 6. Stress-strain behavior of rat tail tendon. Stress response of the tendon is derived from the underlying architecture of the collagenous elements in the tissue. The stress-strain curve in tension is divided into three regions. In the toe region, the slope of the stress-strain curve (the modulus of the tendon) gradually increases with increasing length. This corresponds to a progressively straightening of the waviness in the collagen fibrils. Once all the fibers are fully straightened, the elements in the structural hierarchy are stretched elastically, giving rise to a region of constant modulus on the stress-strain curve. This behavior continues until individual elements at the sub- and microfibrillar levels begin to break or slip in relation to one another. The modulus then decreases with increasing strain until the tendon fails catastrophically. Differences in the hierarchical structure related to the age of the animal are manifested in all these regions (Ref. 7).

This response of the elements of the hierarchical structure of the tendon is reflected in the shape of the stress-strain curve (Figure 6).[7] At small tensile deformations, the stress-strain curve is non-linear which is the case for all soft connective tissue. As the tendon is stretched further, the stress-strain curve becomes linear as a result of progressive straightening of the collagen fibers. This is referred to as the toe region of the curve. All normal physiological loads are confined to this non-linear toe region. When all the fibers are straight, the modulus is constant since the structure is uncrimped. In the linear region, the fully straightened collagen fibers are further pulled elastically. If the load is released, the tendon will entirely recover to its initial crimped morphology. At high strains, the tendon shows yielding and irreversible damage is imparted to the structure since the collagen fibrils begin to disassociate into sub- and micro-fibrils. Localized slippage and voiding between hierarchical levels account for the yielding observed at the macroscopic level. Thus, once the tendon yields, it cannot fully recover to its initial state. The hierarchical design distributes the remote stress locally by imparting damage efficiently throughout the different levels of structure thereby minimizing damage concentrations that could precipitate failure and fracture. This allows the damaged tendon to continue to function in a nearly normal way. The failure of these small structural elements also absorbs a tremendous amount of energy thereby pre-venting catastrophic failure. Ligaments behave in essentially the same manner as tendons except that they connect bone to bone around a joint. As such, their function is to provide a stabilizing force on the joint and prevent unnatural motions from occurring. Examined between crossed polarizers in the optical microscope, collagen fibers in ligaments have the same wavy morphology as in tendons.

2.2 INTERVERTEBRAL DISC

The intervertebral discs are interspersed between the vertebral bones in the spinal column. In this location, the disc is subjected to compressive loads generated on the spine by the weight of the body as well as torsion and bending loads during movement. Another design consideration is impact loading superimposed upon the body weight during activities such as walking or jumping. The disc is composed of two parts: a gelatinous nucleus pulposus containing a matrix of fine collagen fibers, hydrophilic proteoglycan molecules, and up to 88 percent water; and the annulus fibrosis, having concentric cylindrical layers of fibrous collagen arrayed around the nucleus like the layers of an onion skin (Figure 7).[8] The collagen fibers that comprise the annulus are inclined with respect to the spinal axis by the interlamellar angle. In successive lamellae, this angle alternates like a layered lay-up in man-made composite materials.[8] This angle varies with radial distance through the annulus from $\pm 62°$ at the periphery to $\pm 45°$ in the vicinity of the nucleus thus imparting a structurally graduated architecture to the disc. The collagen fibers that make up these layers have the same planar crimped

Figure 7. Radial anatomy of the intervertebral disc. A slice through the anterior of the disc cut parallel to the axis of the spinal column and photographed through the optical microscope in crossed polarized light reveals the complex collagenous architecture. The relationship between the annulus fibrosis, nucleus pulposus, and the cartilage endplates is shown. While the nucleus and cartilage exhibit little optical activity, the lamellae of the annulus are highly active. These lamellae are made up of large collagen fibrils, the ends of which are seen in this view (Ref. 8).

morphology as seen in the tendon (Figure 8).[8] The crimp parameters vary with radial distance through the annulus as well, with the crimp angle increasing from 22° at the periphery to 42° near the nucleus. The unique hierarchical organization of the disc utilizes the superior tensile properties of the collagen fibers of the annulus in resisting compressive forces oriented along the spinal axis as well as torsional and bending forces (Figure 9).[8] Examination of fixed discs in compression reveals that as the disc is compressed, the nucleus pulposus spreads outward laterally causing the lamellae of the annulus to bulge outward. Since the ends of the fibers are anchored to the vertebral bone, they are stretched in tension. This elongation of the crimped fibers again accounts for the observed shape of the stress-strain curve. The stress-strain curve of the disc in compression is similar to that of the tendon in tension, containing toe, linear, and yield regions. While it has not been demonstrated experimentally, one can imagine

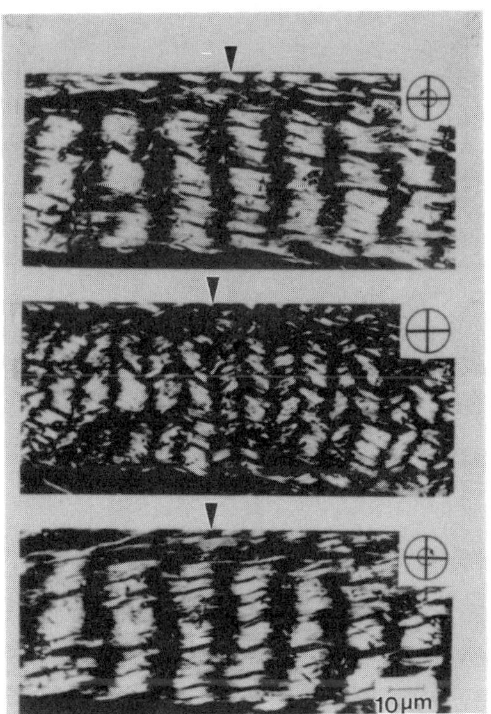

Figure 8. Collagen fibril morphology in the disc. The collagen fibrils that comprise a single lamellae are shown in crossed polarized light at high magnification in the optical microscope. Extinction bands are observed evenly spaced along their length. Upon rotation of the microscope stage, these bands approach one another and eventually merge. The behavior of these fibrils is identical to that of fibrils taken from the tendon and is characteristic of the planar zigzag waveform observed in those tissues (Ref. 8).

that the toe, linear, and yield regions of the stress-strain curve for a disc in compression correspond to the same morphological requirements and changes in the collagen fibers that are present in tendon. The biaxial lay-up of the fibers in the annulus also stabilizes the disc in torsion and bending. In torsion, the collagen fibers in the lamellae oriented in the direction of twist are stretched while the balance are unloaded. In bending, the biaxial lay-up of the fibers prevents the disc from buckling on the flexion side while

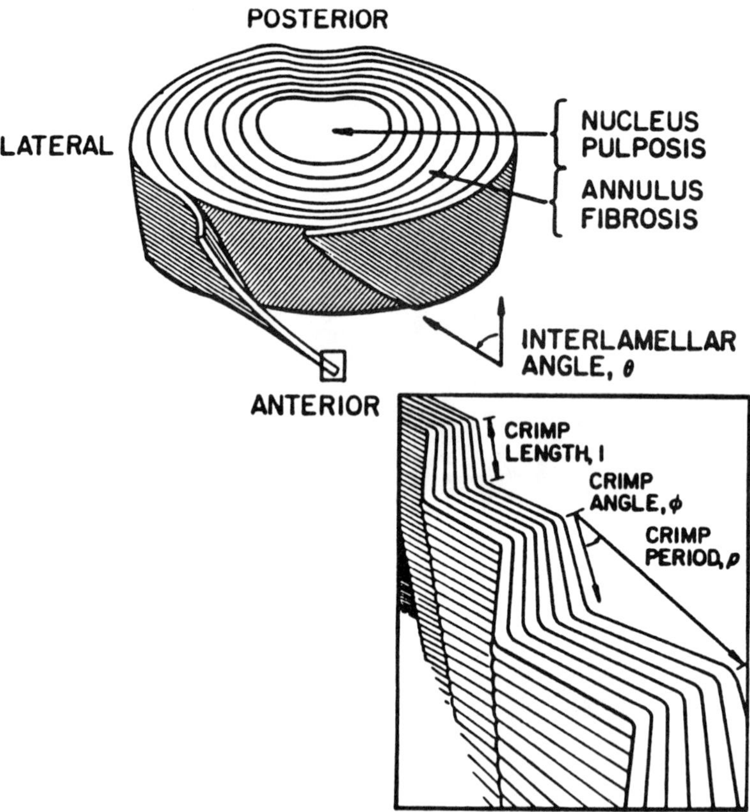

Figure 9. Hierarchical structure of the intervertebral disc. In the intervertebral disc, collagen fibrils are organized into lamellar sheets in the annulus fibrosis which surround a gelatinous and highly hydrated nucleus pulposus. The thickness of lamellae vary with location and are thicker at the anterior and lateral aspects of the disc than at the posterior. Within lamellae, fibrils are parallel and inclined with respect to the axis of the spinal column by an interlamellar angle that alternates in successive lamellae. This angle decreases from the edge of the disc inward. At higher magnification, the fibrils have a planar zigzag waveform. The crimp angle is largest in fibrils close in to the nucleus and decreases toward the periphery. The orientation of the collagen fibrils in the annulus gives the disc strength and stability in tension, bending, and torsional motions. Based upon optical microscope observations of the morphology of the collagen fibrils, the levels of structural hierarchy below the fibrils are assumed to be identical to those of the tendon and intestine (Ref. 8).

resisting excessive bulging on the extension side.

The high water content of the disc plays a vital role in determining the time dependent mechanical response of the disc in compression.[8] If the disc is compressed and held, as it is when the body weight presses the spine during stance, water is squeezed out of the disc. This water loss and the concomitant decrease in disc height is responsible for the "shrinkage" in height that people experience during the course of a normal day's activities. This water is transported out of the disc through cartilage endplates separating the disc from the vertebral bones and into the general circulation. When the load is released, the water is reabsorbed by the disc in the same manner. The stress relaxation and creep responses of the disc in compression are the result of this transport process, rather than the molecular reorganization normally associated with time dependent mechanical phenomena in materials.

2.3 INTESTINE

The intestine is a semipermeable flexible tube in which mechanical break-up of food is accomplished and through which the products of digestion diffuse into the circulation. As this process takes place, the walls of the intestine are stretched in both the longitudinal and the radial or hoop directions. While this requires a design with adequate tensile strength in these two directions, movement of nutrients across the wall of the intestine demands that efficient routes exist for diffusion to take place. The intestine fulfills both these requirements through an innovative hierarchical structure. The intestinal wall is a layered structure with four distinct layers. Two layers contain smooth muscle cells oriented in the longitudinal and circumferential directions to macerate food particles; another is a covering that isolates the intestine from the other organs in the body cavity.

The predominant structural member of the intestine is the submucosa, which contains the majority of the collagen fibrils. Fibrils are arranged in layers oriented biaxially around the lumen. In successive layers the fibrils are oriented at ±60° to one another forming helices of opposite sense arrayed at ±30° to the longitudinal axis winding around the intestine (Figure 10).[9] This biaxial winding of the collagen fibrils prevents

Figure 10. The biaxial crimped layered structure of collagen fibers in the intestine. Collagen fibers in the intestine have the same planar crimped morphology observed in the tendon. The parameters of the waveforms (crimp angles, period) are different, however, with the greater extensibility required in the intestine being reflected in a large crimp angle (Ref. 9). *(See also **Color Plate** 2.)*

buckling of the intestine as it winds through the body cavity while the multiple layers of helically wound fibers gives maximum resistance to internal bursting pressures. Within each layer the collagen fibrils are planar crimped as in the tendon. The hierarchical structure of the intestine below the level of the fibrils is identical to that of the tendon (Figure 11).[10]

The crimped morphology in conjunction with the biaxial orientation of the collagen fibers gives rise to the shape of the stress-strain curve in tension and to the anisotropic nature of the mechanical response. At low tensile strains, a similar toe region exists as in the tendon. However, in the intestine, there are two deformation mechanisms at

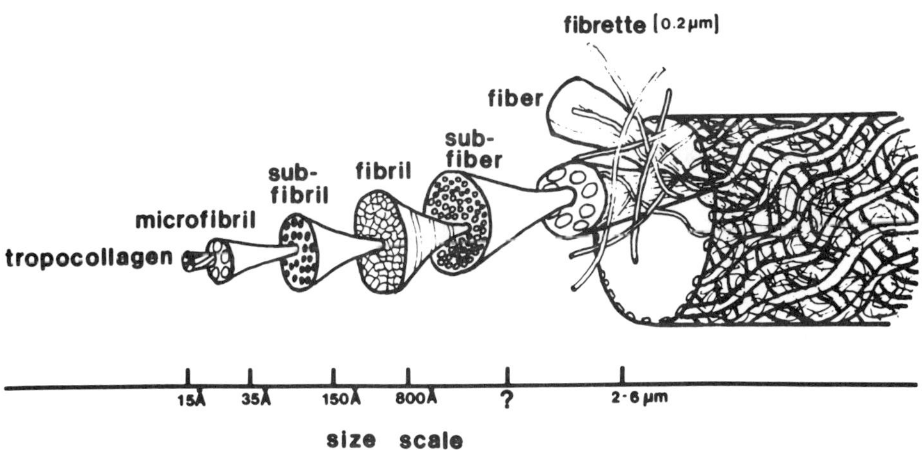

Figure 11. Hierarchical structure of the intestine. The hierarchical organization of the intestine is identical to that of the tendon from the molecular through the fibrillar level. The two tissues differ only in how the collagen fibrils are arranged into the highest levels of structure. Note that in the intestine no fascicles are present. Instead the collagen fibrils aggregate into fibers which are then wound around the intestine in a helical fashion (Ref. 10).

work due to the requirement of simultaneous biaxial expansion. A limited reorientation of the fibers, as evidenced by a decrease in the angle between fibers in adjacent layers, takes place along with the progressive straightening of the crimp in the fibers. The biaxial lay-up of the fibers also causes an anisotropic mechanical response in tension. If the intestine is stretched along different orientations with respect to the longitudinal axis, different stress-strain responses are observed (Figure 12).[11] Intestine is the stiffest and least extensible at $\pm 30°$ to longitudinal, in alignment with the fibers in one set of layers, while the greatest extensibility is at 90° to the longitudinal axis. This is suited to the function of the intestine which necessitates changes in the diameter of the intestine

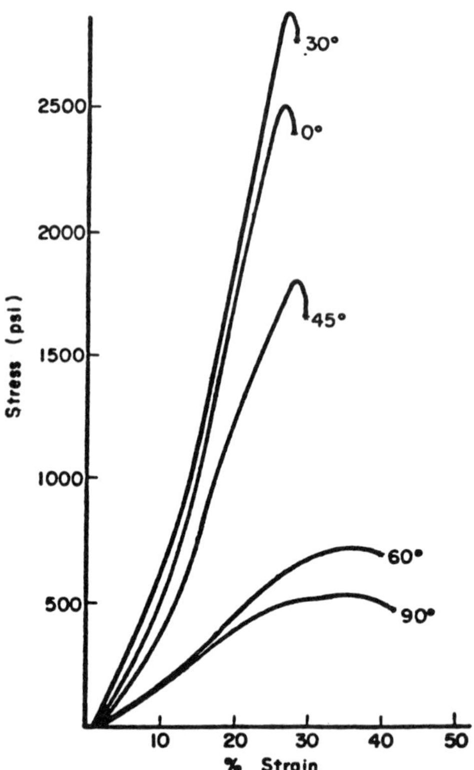

Figure 12. Stress-strain behavior of the intestine in tension. The tensile properties of the intestine vary with orientation as would be expected in a tissue made up of biaxially oriented fibers. The greatest modulus and least extensibility occur when the intestine is stretched parallel to the fibers in the layers (30°). The properties along the longitudinal direction (0°) are only slightly less than in the fiber direction. The greatest extensibility and least modulus are found perpendicular to the long axis in the hoop direction (90°). These properties represent a structural adaptation to the function demands of the intestine. Digestion of food requires the ability to accommodate changes in the diameter of the intestine as food passes through and for rapid diffusion of nutrients through a permeable wall. At the same time, changes in the length of the intestine must be minimized in order for the peristaltic action of the smooth muscle to operate efficiently. The arrangement of collagen fibrils in the intestine enables these demands to be met (Ref. 11).

as food passes rather than changes in length. Once fully straightened, the collagen fibers are strong and inextensible, effectively resisting bursting.

Virtually all connective tissues, whether soft or hard, have hierarchical structural designs arranged at discrete levels of structure. For example, the cornea has mechanical requirements similar to that of the intestine. Collagen fibers must contain the pressures generated internally in the eye by the vitreous. However, its distinguishing characteristic is its transparency to light in the visible region of the spectrum. Both these functions are accomplished using a layered structure as in the intestine. The principal structural layer is the stroma which makes up ninety percent of the corneal thickness. The stroma is a layered structure with approximately 200 lamellae between 2 and 3 mm thick. Within these lamellae, the collagen fibrils are parallel to the surface of the cornea and extend entirely across it. Within lamellae, the fibrils are parallel to one another while in successive lamellae, fibrils make large angles to one another (Figure 13). This lamellar structure and cross-ply orientation provide enhanced reinforcement for the eye when compared to an isotropic arrangement of collagen fibrils. The arrow distribution of fibril thicknesses, uniform packing scheme, and the highly ordered arrangement of the fibrils in the lamellae permits the stroma to perform its mechanical function while passing visible light undisturbed.

3.0 GENERAL COMMENTS

It is tempting to suggest that all connective tissues follow at least three "Laws for Complex Assemblies." Starting with a very similar macromolecular design (the fibrous protein, collagen, and the ground substance, mucopolysaccharide), the hierarchical structures are assembled into distinct and totally different systems both in their morphology and their function.

First, the macromolecules associate into discrete levels of organizations. Usually these are in the form of fibrils which are themselves composed of smaller subfibrils and microfibrils. The fibrils themselves are composed of smaller subfibrils and microfibrils. The fibrils are subsequently arranged in layered structures reflecting the specific

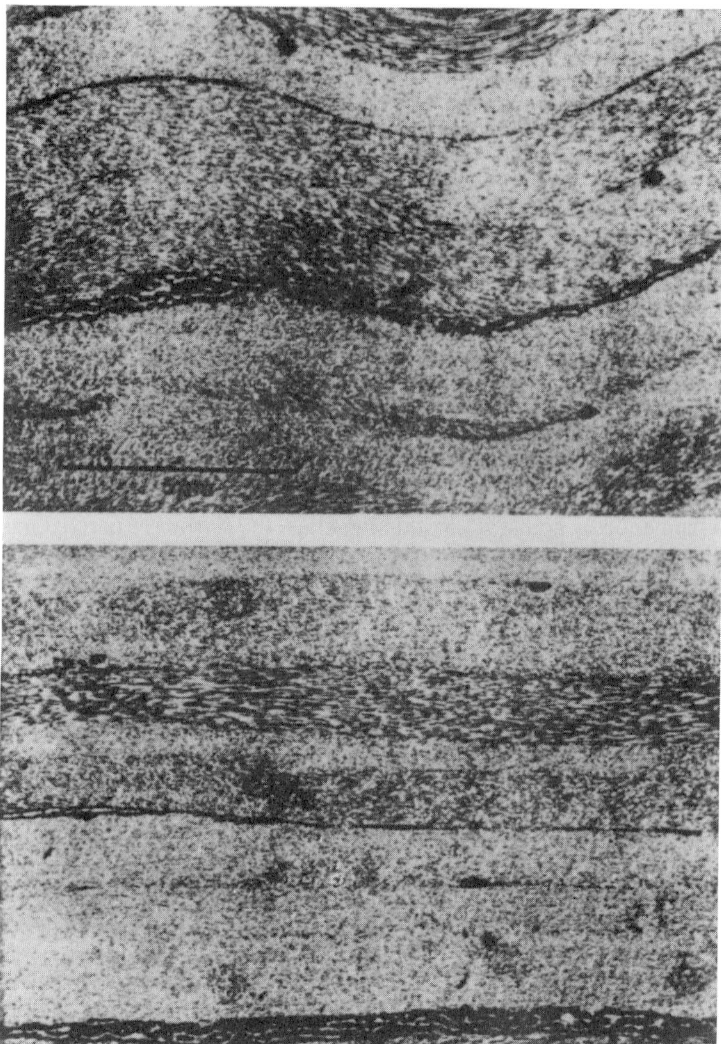

Figure 13. Collagen fiber lay-up in the cornea. Electron micrographs of the central rabbit cornea show the wavy morphology of the collagen fibers extending through the thickness of the stroma. Note also the layered structure of the stroma, the uniform fiber diameter, and the variation in the orientation of the fibers in each layer (Ref. 12).

functional requirements of the overall composite system. The minimum number of discrete levels or scales observed thus far in biocomposites is four. That is, structural levels at the molecular, at the nano, at the micro, and at the macro scale are the minimum required components within an ordered hierarchical biocomposite system.

Second, these levels are held together by highly specific interactions between surfaces. Considerable evidence exists for surface to surface interactions due to intermolecular bonding at specific active sites, and for epitactic arrangements of a crystallographic nature. Whatever the nature of the interfacial mechanism, strong interfacial interactions are required having chemical and physical specificity.

Third, the highly interacting fibers and layers are organized into an oriented hierarchical composite system which is designed to meet a spectrum of property or function requirements. This law of architecture, stipulates that as the complexity of the overall system and its uses increases, the system is capable of a higher degree of adaption in difficult and unexpected environments. The so-called "intelligent composite system" results from a complex architectural arrangement which is designed to serve highly specific functions.

At the very least, the analysis of complex behavior in natural polymeric systems in terms of hierarchies is an approach that interrelates our understanding of structure at various scales. Such an approach may prove invaluable in the design of new advanced polymers. Structural hierarchy is more than a convenient vehicle for description and analysis. Important and difficult questions that remain to be addressed include the physical and chemical factors that give rise to relatively discrete levels of structure and the relations that govern their scaling.[2]

4.0 REFERENCES

1. E. Baer, *Scientific Amer.,* **254**, 179 (1986).

2. E. Baer, A. Hiltner, and H. D. Keith, *Science*, **235**, 1015 (1987).

3. J. W. Smith, *Nature*, **254**, 157 (1968).

4. E. Baer, J. Diamant, M. Litt, R. G. C. Arridge, and A. Keller, *Proc. Royal Soc.*, **B180**, 293 (1972).

5. J. J. Cassidy, A. Hiltner, and E. Baer, *Ann. Rev. Mater. Sci.*, **15**, 455 (1985).

6. E. Baer, S. Torp, and B. Friedman, in: *Structure of Fibrous Biopolymer*; E. D. T. Atkin and A. Keller (eds.) (Butterworth, London, 1975, p. 223);
S. Trop, E. Baer, and B. Friedman, Proc. Colston Conf., 226 (1974).

7. E. Baer, J. Kastelic, and A. Galeski, *J. Connect. Tissue Res.,* **6**, 11 (1978);
E. Baer, J. Kastelic, and I. Palley, *J. Biomech.*, **13**, 887 (1980).

8. A. Hiltner, J. J. Cassidy, and E. Baer, *J. Connect. Tissue Res.*, **23**, 75 (1989).

9. A. Hiltner, J. W. Orberg, and E. Baer, *J. Connect. Tissue Res.*, **11**, 285 (1983).

10. J. W. Orberg, E. Baer, and A. Hiltner, *Connect. Tissue Res.,* **9**, 187 (1982).

11. K. Fackler, L. Klein, and A. Hiltner, *J. Microsc.,* **124**, 305 (1981).

12. R. L. McCally, C. B. Bargeron, W. R. Green, and R. A. Farrell, *Exp. Eye Res.,* **37**, 548 (1983).

NACRE: PROPERTIES, CRYSTALLOGRAPHY, MORPHOLOGY, AND FORMATION

Mehmet Sarikaya,* Jun Liu,* and Ilhan A. Aksay§

*Department of Materials Science and Engineering
University of Washington, Seattle, Washington 98195; and
§Department of Chemical Engineering, and Princeton Materials Institute
Princeton University, Princeton, New Jersey 08544-5263, USA

Biological hard tissues are composite materials incorporating both inorganic component (e.g., phosphates and carbonates) and organic component (macromolecular structural units including proteins and polysaccharides). These materials have excellent physical properties mainly because of their hierarchically ordered structures through the dimensional scale from molecular to submeter. These composites are a source of inspiration for design and processing of synthetic materials based on both their structure and processing. In this chapter, we give a general overview of some biological hard and stiff tissues in biomimetics research. We specifically discuss mechanical properties of nacre section of mollusk shells which is a composite of $CaCO_3$ platelets surrounded by an organic matrix. The current understanding of micro- and nanostructures of $CaCO_3$ and the organic matrix, crystallographic relationship between them, and finally the morphology and shape formation of the shell are discussed.

1.0 INTRODUCTION

A central issue in materials science is to develop materials for specific applications through a precise control of their structural features in order to achieve a desired set of properties.[1] Ideally, the structural features are to be controlled from molecular to macroscopic dimensions. In biologically produced composites, nanocomposite structures themselves are the building blocks for larger scale composite structures as a hierarchical "materials system" builds up from nanometer length scale upward. In synthetic composites, a typical length scale for the smallest level of hierarchy is in the micron range. Exceptions to this rule are found in a few synthetic composites where structural hierarchy is pushed into the nanometer range not necessarily by deliberate design but by efforts focused on the improvement of properties through trial-and-error methods. A classic example is Fe-C steel, in which metastable equilibrium structures can be produced through thermal and thermo-mechanical treatments. Countless different steels have been produced since the eighteenth-century Industrial Revolution, and many are still in use today (in fact, tough and strong steels were produced for hundreds of years earlier as materials for armor, for example Damascus and samurai swords, and simple instruments).[2]

1.1 NANOCOMPOSITE MATERIALS

Recent research has shown that materials with controlled structural variations at the nanometer scale (i.e., the size range from 1 to 100 nm, hence the name nanostructured or mesoscopic)[3] exhibit unprecedented physical, electronic, optical, and magnetic properties. Fundamental reasons for property improvements are still not well understood. These nanostructured materials are a rather natural outcome of the extensive work and increased understanding of the structural effects on many Fe-C[4] and Al alloys,[5] cemented carbides,[6] and superalloys[7] throughout this century, especially right before and after the World War II. In each case, toughness increases without a loss of strength and, in fact, concurrent with an increase in hardness. This is because of controlled precipitation of nanometer-sized metastable precipitates in the matrix through tempering.[4] In laminated single or multiphase composites, elastic modulus (or hardness) increases as the thickness of each laminate is decreased below 10 nm, the so-called supermodulus effect.[8] High temperature superconductors, such as $YBa_2Cu_3O_{7-x}$, may be considered laminated composites with alternating oxide and metallic layers (Cu-O and Y or Ba) within a dimension of little more than one nanometer.[9] In non-linear optical materials, such as Lead Zirconate Titanate,[10] controlled variation of the structure at the nanometer scale changes the second- and third-order parameters and creates non-linear optical effects and novel electrical properties.[11]

Although some current synthetic techniques can achieve limited controlled composite structures, it is still difficult to produce materials in practical and energy efficient ways and in large quantities with desirable interface properties and with a high degree of control of the lattice and defect structures at the nanometer scale. Some of these problems may be circumvented as in the synthesis of semiconductive multilayer quantum-well structures, such as Si-Ge and GaAs-GaAlAs, with the use of ion- or molecular-beam techniques to control structural development truly at the atomic scale.[12,13] Even here, the size of the material produced is limited and the processing is performed under the strictest conditions of temperature, pressure, and environment. As a result, these techniques do not have widespread structural application except in areas that require small and critical components. Manufacturing of almost all large technological materials presents more difficulties. First, high temperature processing and consequent energy inefficiency are typical. Second, structural design is usually accomplished at a single length scale, with possible full control, and over several scales in synthesis, with limited control. Third, these materials are typically processed for one particular property and thus are not multifunctional. Finally, technological materials do not actively respond to external stimuli and are not damage tolerant.[14-17]

Biological systems are a rich source of inspiration for design and processing concepts for developing novel synthetic materials where structural control is established over a wide length scale.[18-22] From a materials scientist's perspective, some key features of the biological materials are: (i) synthesis is accomplished at low temperatures (<100°C) under ambient conditions in aqueous environments; (ii) structural make-up is controlled at a continuous length scale from molecule to final tissue in a hierarchical manner; and (iii) the resultant composites are multifunctional and self-healing provided that the cellular activity is maintained. The approach used by organisms in processing materials is, in many ways, more controlled than synthetic methods.[21-24] The dynamics of these systems allow the collection and transport of the raw constituents,[25] self-assembly,[26] nucleation, configuration, and growth of new structures,[27,28] and the repair and replacement of old or damaged components.[29]

We divide design approaches based on biological systems into two categories (Figure 1).[30] First, by investigating the structures of biomaterials at all possible scales of spatial resolution, the fundamentals of their unique structural designs can be deduced and then mimicked by techniques currently available to materials scientists - an approach we refer to as *biomimicking*. Second, by mastering the molecular synthesis and processing mechanisms of biomaterials, these hitherto unknown methodologies may be applied to produce new technological materials, superior to those presently available - an approach we call *bioduplication*. The bioduplication approach is much more complex than biomimicking and will require a long-term commitment (probably decades of research), not only to learn the intricacies of bioprocessing used by organisms but also to develop new strategies to process materials synthetically from the molecular level up with the same size, shape, complexity, and multifunctionality as the biocomposites. The biomimicking approach, although by no means simple, will require a shorter time commitment (ten years of research or less). Biomimicking involves exploring the structures of biomaterials, correlating their multifunctional properties with the specific structural features. This chapter focuses only on the biomimicking approach.

We describe some of our findings on structural design of the nacre of abalone shell and answer several questions about this unique, but relatively simple, composite. In particular, we focus on the organization of the inorganic component, its possible

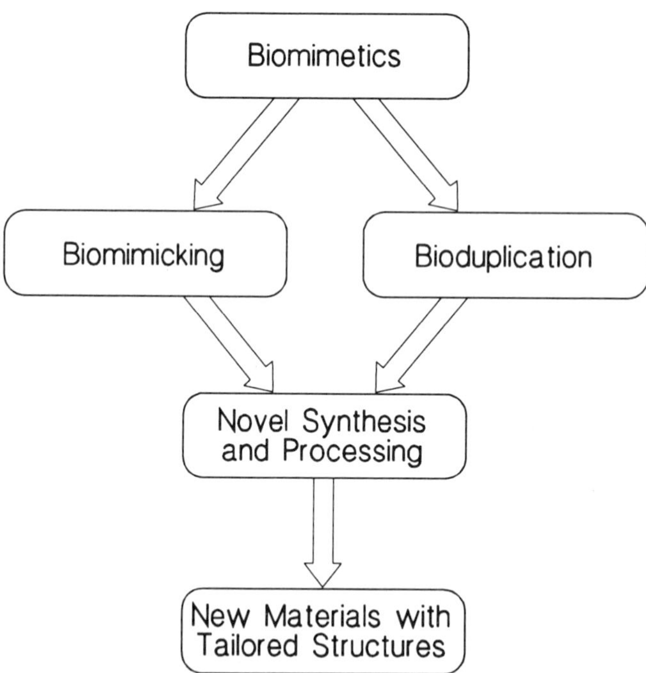

Figure 1. Schematic description of methods of materials design based on biomimetics.

structural relation to the organic matrix, and the growth mechanism of the shell. Before describing a detailed structure of nacre, certain unique aspects of some of the other biological composites are discussed. Our goal is to obtain lessons and design guidelines for the synthesis of future technological materials via biomimetics which, to us, is a natural extension of the materials revolution that is taking place today and is expected to continue well into the next century.

1.2 AN OVERVIEW OF MICRO- AND NANOSTRUCTURAL DESIGN IN BIOLOGICAL HARD TISSUES

Biological materials that are of interest in technological applications are mostly hard and stiff tissues and small particles. These structures are composed of an organic matrix (mostly proteins and polysaccharides) with the inorganic phase interspersed throughout. The formation, morphological development, and crystallography of the inorganic particles are assumed to be controlled by the organic matrix, a hypothesis which has not been indisputably established.[21,31] In the case of stiff tissues, such as insect cuticles,[32] all the components of the composite are organic macromolecules in which the matrix is usually composed of proteins. The stiffness comes from fibrillar polysaccharides, such as collagen and chitin, organized at the nano-, micro-, or higher-scales.[33] Some lower organisms, such as bacteria and algae, produce fine inorganic particles with unique physical properties.[34] As listed in Table-I with their corresponding properties, biological composite materials include all organic components,[35-40] such as spiders' webs,[35] mucus,[36] and insect cuticles;[32,33] inorganic-organic composites, including seashells,[41] teeth,[42] and bones;[43] ceramic composites, such as sea-urchin teeth[44] and spines;[45] and inorganic ultrafine particles, such as magnetic[34,46] and semiconducting[47] particles produced by bacteria and algae, respectively. In some cases, byproducts and enzymes,[48] proteins,[49] and other macromolecules[50] have chemical or physical properties superior to their synthetic counterparts.

Small inorganic particles of biological origin offer analogies with synthetic nanoparticles[11] and mesoscopic systems.[51] There are many organisms that produce ultrafine inorganic particles that perform various functions.[34] One notable example of this is iron clusters that form at the center of ferritin molecular cages (or vesicles) in organisms.[52] In some cases, metal clusters are accumulated as foreign entities that might

Table-I. Categorization of various biological composites, their micro-/nanodesign, and physical properties

Material/composite	Example	Scale	Properties
Small particles	bacterial algal	N N	magnetic, electronic, optical
Ceramic/ceramic	sea-urchin	B	mechanical (wear resistant)
Ceramic/polymer	mollusk bone dentin	N, H N, H N, H	mechanical (tough, strong), ferroelastic
Polymer/polymer			
laminated	cuticle	N, H	mechanical, optical
fiber/matrix	tendon	N, H	mechanical, ferroelastic
fiber/fiber	silk	N	mechanical (tensile)
Liquid crystalline matrix	mucus	N	rheological

N: nanometer; M: micrometer; B: both nanometer and micrometer; H: hierarchical

otherwise be harmful to the host organism, such as CdS in algae.[47] Another example is ultrafine magnetic particles that are found in bacteria.[34,46,53,54] Some species of bacteria that live anaerobically in freshwater and salt swamps move about to seek oxygen and food by a mechanism that makes use of a string of magnetic particles (Fe_3O_4 or Fe_3S_4) as a compass.[46] In *Aquaspirillum magnetotacticum*,[55] for example, each bacterium has about 20-25 particles that are oriented along a string with their magnetization axis along the long axis of the bacterium (Figure 2). Bacteria can move either forward or backward, depending on their configuration with respect to the Earth's magnetic field. In terms of biomimetic applications, some of the significant materials characteristics of these particles are: (i) they are single crystalline (no lattice defects, such as dislocations, twins, or stacking faults),[54,56,57] (ii) they have a uniform particle size of about 500-600 Å, and, thus, are in the single domain (superparamagnetic) region,[55,58] (iii) particle shape is species specific and can be dodecahedral, cubo-octahedral, or hexagonal,[56] (iv) they are aligned as a single string of particles (and occasionally as double strings)[54,59] (v) they form in biological sacks, called magnetosomes.[54] In S-rich regions, some species are known to form an isomorphic form of magnetite, i.e., Fe_3S_4.[60]

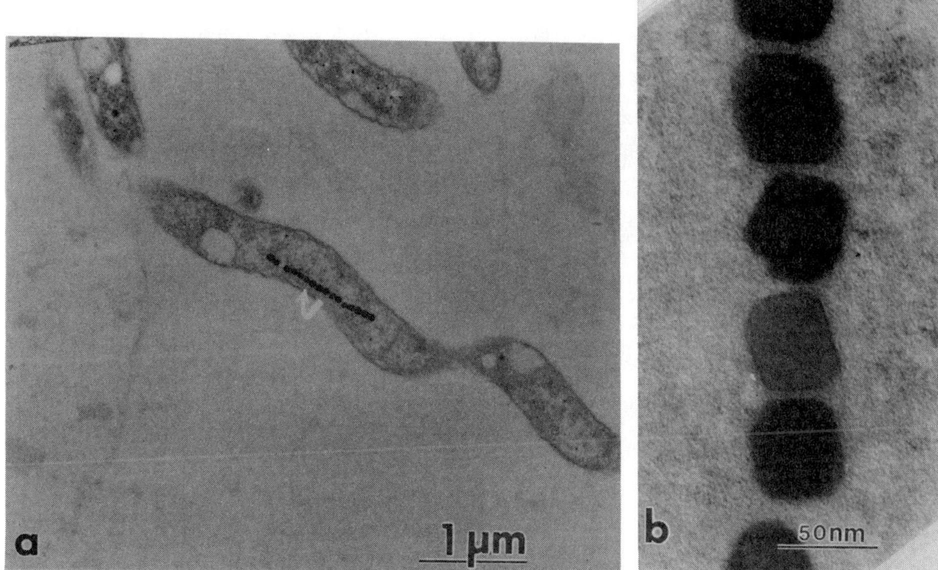

Figure 2. TEM images of *A. magnetotacticum* show a string of fine Fe_3O_4 particles.

The formation of magnetic particles within magnetosome membranes is of great interest especially with respect to forming small synthetic particles under closely controlled synthesis conditions.[61] Small magnetic particles can be formed synthetically in several ways including solution precipitation from precursors, in microemulsions, and using vesicles.[62] In each of these cases, however, the particles formed are nonuniform, often are not fully crystalline, are compositionally nonhomogeneous, and, more importantly, are in an agglomerated state which imposes problems in processing.[63] Processing of small magnetic and other inorganic ultrafine particles via biological routes, therefore, promises advantages in terms of controlling synthesis and morphological properties. The most critical point in the understanding of particle formation in magnetosomes is mechanism(s) by which organic matrix allows the nucleation and controls the growth of these particles. The investigation of structure and composition of the magnetosome membrane and its protein organization, therefore, is important for understanding transport of ions through the membrane and the early stages of particle formation. Current knowledge of the membrane demonstrates that it is a bilayer and is likely to contain proteins that are found in the outer membrane of the bacteria.[64] Other questions involve details of the early stage of formation of particles (amorphous or crystalline and, possibly, in the hydrated form), selection of their chemistry (e.g., Fe_3O_4 vs Fe_3S_4), the control of their growth, and finally, factors that affect particle morphology, as discussed by Frankel and Bazylinski in this book.[65]

While only a small particle may be wholly inorganic in bacterial magnetite, in some organisms, such as echinoderms[66,67] skeletal components display unique composite structures also involving mostly inorganic phases. In biocomposites, such as sea shells and bones, it is assumed that an organic phase is usually associated with an inorganic phase in an easily recognizable way, with each phase in close proximity to the other.[20,21] In sea-urchin skeletal units inorganic phase appears to be present alone and constitutes the overall component of the skeleton. (In these systems, organic matrix may cover the overall skeletal unit as a sheath (teeth) that may control the growth, or may be present within holes of the spongy-structured inorganic phase (spine) which is dissolved during sample preparation for observation.) These biocomposites, therefore, also provide potential examples in biomimetic applications for ceramic/ceramic composites. In the body and the spine of the sea-urchin [Figure 3(a)], the mineral is a calcitic single crystal forming an intricate architectural design.[68] The most interesting among the structural composites in sea-urchin are its teeth.[69] Five pieces of teeth in the lower center of the body are used to scrape food from the surface of rocks. A cross-section at the cutting edge of a tooth is composed of a matrix of amorphous $CaCO_3$ with crystalline calcitic $CaCO_3$ fibers embedded with their long axes perpendicular to the cutting surface to increase the wear resistance[69,70] of the tooth [Figure 3(b)]. Irrespective of sea-urchin teeth, this design has been in use in synthetic fiber-reinforced composites.[71]

Recent studies reported that organic macromolecules may be occluded within certain sea-urchin skeletal units.[67] With respect to biomimetics, the major questions involve the presence, types, and spatial distribution of organic macromolecules that constitute less than 1 vol. % of the structure. If there exist occluded proteins, or their fractions, then these hard tissues may be regarded as molecular composites, i.e., analogs of nanocomposites.[3] Investigation of sea-urchin at high spatial resolution (in imaging) and elemental resolution (in spectroscopy) is expected to provide better insights into the understanding of these unique structures.

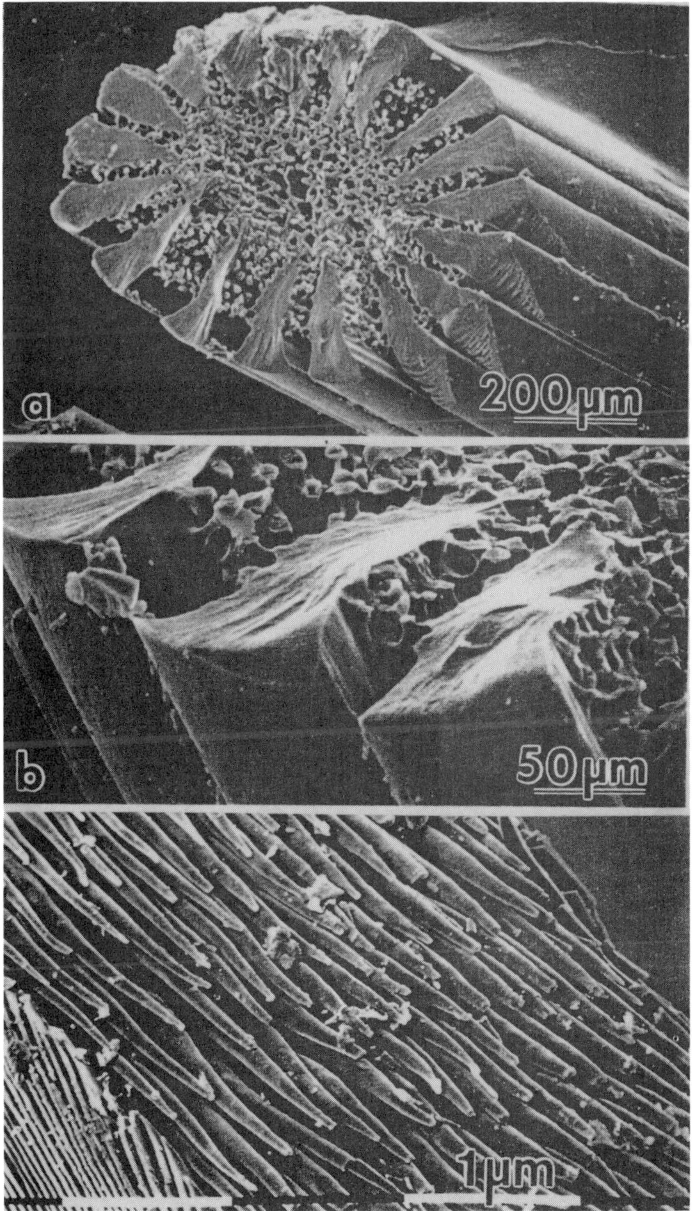

Figure 3. SEM images of a sea-urchin (a,b) spine, showing its intricate structure, and (c) tooth, displaying its composite structure consisting of calcite fibers embedded in an amorphous $CaCO_3$ matrix.

Organic/organic biological composites constitute numerous stiff biological tissues, composites of fibrous organic components embedded in a soft organic matrix, that are analogs of fiber- or particle-reinforced polymeric composites.[71] Tendon,[38,72] which connects muscle and bone, is a classic example. It has six discrete levels of structures organized in a hierarchical manner from molecular to centimeter-scale. Silk, found in cocoons of silk moths and webs of spiders, is another structural material.[35,73i] These unique structures consist of silk fibroin proteins (pleated sheets) organized in a liquid crystalline fashion in an amorphous protein matrix. These are designed, by the organisms, to withstand stresses much higher than those encountered by high tensile strength-metallic or polymeric fibers (see also the paper by Gosline et al.).[73ii]

One of the major classes of organic biocomposites is cuticles.[32,74] Arthropods, such as insects, crustaceans, spiders, millipedes, and others, are evolutionarily very successful, probably because their cuticles cover their bodies from "head to toe." The cuticle, therefore, is the skeleton, for example, of an insect. Its structure resembles that of fiber-reinforced polymer matrix; the fibers are chitin (polysaccharides), and the matrix consists mostly of proteins.[74,75] The composite has a sheet structure in which chitin fibers are arranged in layers. In each layer, the fibers are oriented parallel to each other.[32,33,75] In successive layers, there is a rotation of these parallel fibers only a few degrees, i.e., helicoidal (discussed by Gunderson and Schiavone in this book and references therein).[76] The unique structure of the insect cuticle, as well as cuticles of other classes, may serve as an example for the design of composite materials in which all the components are polymers.

Exoskeletons, in addition to serving as protective "armor" for insects, also form intricate surface structures which give optical effects. For example, in butterfly wings, although the origin of color for the most part is pigmentary, some colors such as blues and violets come from scattering of light from highly ordered surface structures.[77] The structural details on surfaces can take an intricate and ordered combination of layers, scales, and ridges which are arranged to produce optical effects through the interference of light. Coloring through structural variations is an area of study in biomimetics. Furthermore, pigmentation itself is also an area of materials science interest, involving questions such as the origin of pigmentation, the nature of pigments, their size, distribution, and composition within the matrix, and their coloring effects.

In cuticles, there are both biomimicking and bioduplication possibilities for future materials formation. In the case of biomimicking, questions involve the types of structural units at the nano- and microscales, and their spatial distribution. There are already synthetic polymers that may be used as the matrix (such as diblock copolymers),[78] as well as fibrous polymers[79] that can be used as the stiff component. In the case of bioduplication, some subjects of future investigation are composition and spatial distribution of proteins in the matrix, their possible liquid crystalline order, and their synthesis. The nature of chitin, its molecular structure and composition, its structural relationship with the protein matrix, and overall hierarchy of the structure of cuticle will be further areas of biomimetic interest.

The examples of biological composites that contain both ceramic and biomacromolecular phase(s) vary in their form. They are mainly encountered as structural materials in skeletons or as protective covers over the bodies of organisms.[20,21] These include bones[80] in vertebrates, teeth[81] in fish and mammals, and shells in mollusks[20,21,82] One of these biological composites, namely the nacre section of mollusks, is discussed in detail in the rest of the chapter.

1.3 A BACKGROUND ON NACRE STRUCTURE

Nacre is a biogenic composite of a ceramic phase and macromolecules.[30,83-85] Its significance as a structural material stems from its excellent mechanical properties, such as fracture toughness and strength, which are comparable or better at room temperature than those of some high-technology structural ceramics.[83-85] This result is the major driving force in our quest for producing ceramic-based composites (cermets and cerpolys) with mechanical properties better than the existing ones.[30,85]

The major component of nacre is $CaCO_3$, a material with limited engineering value in the bulk form for structural applications (although it is heavily used as a filler in the powder form in cements, papers, paints, and plastics).[86] This unique structure is composed of alternating nanometer-scale laminated layers of thin biomacromolecular matrix and $CaCO_3$ (aragonite) platelets, all highly organized to produce an excellent multifunctional material (armor) for the organism.[83-85] Some of the other known facts about nacre include:[83-85,87-89] (i) the overall shell composite is more than 95 vol. %

inorganic material; (ii) the composite has a brick and mortar architecture with aragonite forming thin, hexagonally shaped multi-edged bricks[30,90] and an organic matrix that is a composite of proteins and polysaccharides (Figure 4);[90] and (iii) the inorganic phase consists of crystallographically highly oriented aragonitic platelets.[30,90i] If the desired synthetically laminated materials based on nacre are to be produced through biomimicking and bioduplication, further knowledge is needed about this unique structure. The areas of investigation may be divided into several categories: *(i) micro- and nanostructural variations*, including: (a) lattice and interface defect structures within the aragonitic phase, (b) composition and structural organization of the organic matrix, (c) interface structural and compositional relationship between aragonite and the organic matrix; *(ii) structure-property relationships*, including: (a) cooperative and separate structural responses of phases (aragonite, organic matrix, and interfaces) under various mechanical stresses, (b) toughening, strengthening, and hardening mechanisms, (c) size, lamination, and organizational effects (nanocomposite behavior and source(s) of multifunctionality); *(iii) biomineralization*, including: (a) identification of the active units within the organic matrix (nucleator proteins), (b) mechanism of self-assembly of the organic macromolecules (also of construction of framework macromolecules), and (c) mechanisms of nucleation and growth; *(iv) shape formation (morphogenesis)* including: (a) hierarchy in structural units and (b) growth mechanism of the shell (based on geometry, crystallography, and size); *(v) biomimicking*, including (a) detailed structural build-up of the shell from the molecular level to macro-scale and (b) its mimicry through novel materials processing strategies; and *(vi) bioduplication*, including (a) separation, purification, compositional (amino acid) and structural identification of each of the macromolecular units in the organic matrix, (b) cloning, reproduction, and self-assembly of structural macromolecules, and (c) processing new materials by using these macromolecules based on both biomineralization and morphogenesis principles. Answers to these questions will shed more light not only on the nanocomposite behavior of nacre but also on the mechanisms of its formation and particularly the degree of control that the organism has over the growth of this highly ordered biocomposite, and their use in the processing of future biomimetic materials.

Although discussed in our earlier papers, for completeness of this book, we begin with a summary analysis of the mechanica! properties of nacre. A set of criteria is developed for the structural design of synthetic laminated composites. This is followed by a

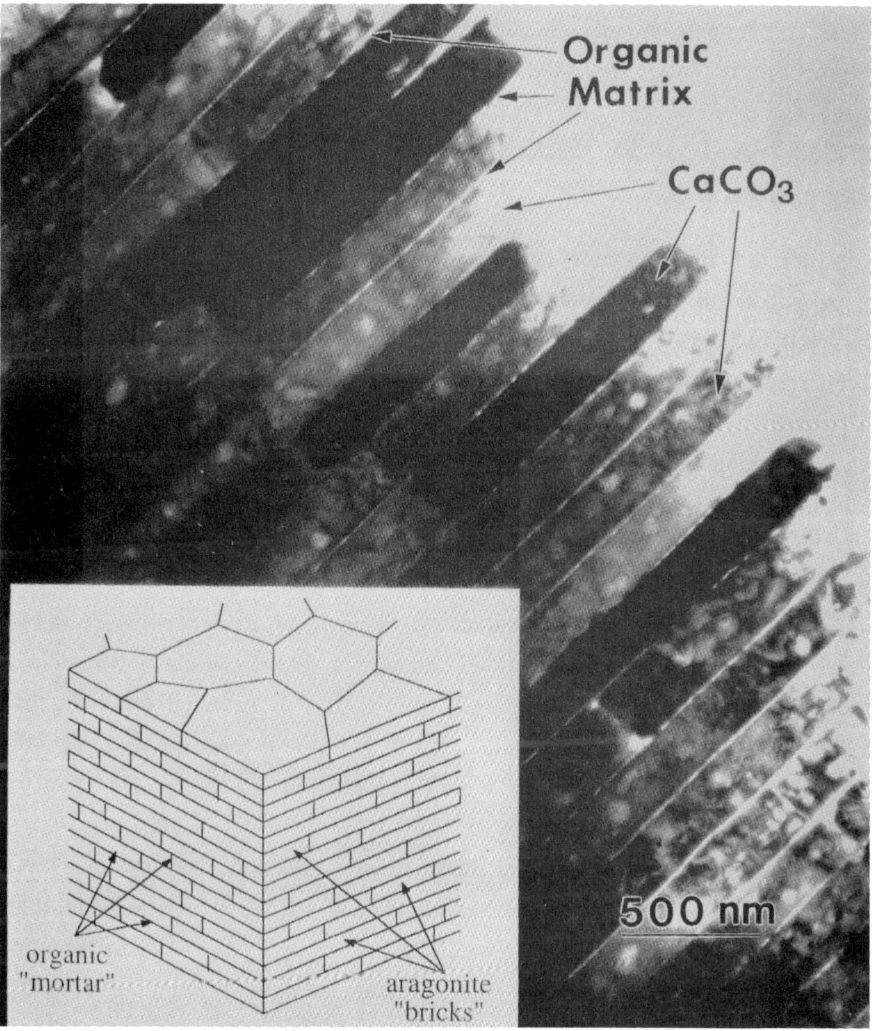

Figure 4. A cross-sectional image of nacre (edge-on configuration) of red abalone. The inset is a schematic three-dimensional view of the structure.

description of the current understanding of the structure of nacre in terms of morphology, composition, and crystallography. Its hierarchy is also discussed in both the hard and soft tissues that lead to a possible structural relationship between the organic and inorganic components. Finally, an assessment is made of current knowledge, with possible directions, for future studies.

2.0 MECHANICAL PROPERTIES OF NACRE: A NANOLAMINATED CERAMIC-POLYMER COMPOSITE

Nacre structure is found in many families of mollusks, including abalone (*Haliotidae*) of the gastropod family, cephalopods such as nautilus (*Nautilus pompilius*), bivalves such as pearl oysters (*Pinctada)*, and blue mussels (*Mytilus edulis Linne*).[91-94] A transverse cross-section of each of these shells displays two types of microstructures: an outer prismatic layer (calcite) and inner nacreous layer (aragonite). We focus our work on the structure and properties of the nacreous layer since this is the part of the shell that displays an excellent combination of mechanical properties as a result of its highly ordered laminated structure. We used mostly red abalone (*Haliotis rufescens*) because in this species the diameter and thickness of the shell containing the nacre are large enough to perform standard mechanical tests.[95] This allows a direct comparison of nacre with engineering structural ceramics and ceramic-based materials. Black-lipped pearl oyster (*Pinctada margaritifera*) shells were also used. Although relatively thin (1 to 5 mm), these shells are much more uniform through the thickness and relatively flat, and, therefore, exhibit less scatter in mechanical tests.

Red abalone samples were collected in Baja California, Mexico. The specimens were estimated to be 10 to 12 years old, had shell diameters of 25 to 30 cm, and shell thickness of about 1.5 cm, including the prismatic and nacreous layers. Mechanical properties in terms of fracture toughness (K_{IC}) and fracture strength (σ_F) in three-point straight notched and four-point bend bars, respectively, were evaluated in the transverse direction - perpendicular to the laminated layers through which a crack is normally expected to propagate in the shell of the mollusk in its natural environment.[96] Fracture toughness and fracture strength of nacre of both abalone and pinctada, and those of some of the well-known ceramics and ceramic-based composites (cermets) are plotted in Figure 5. The average K_{IC} and σ_F values of nacre are approximately 20-30 times that of geologically produced monolithic $CaCO_3$.[84,85]

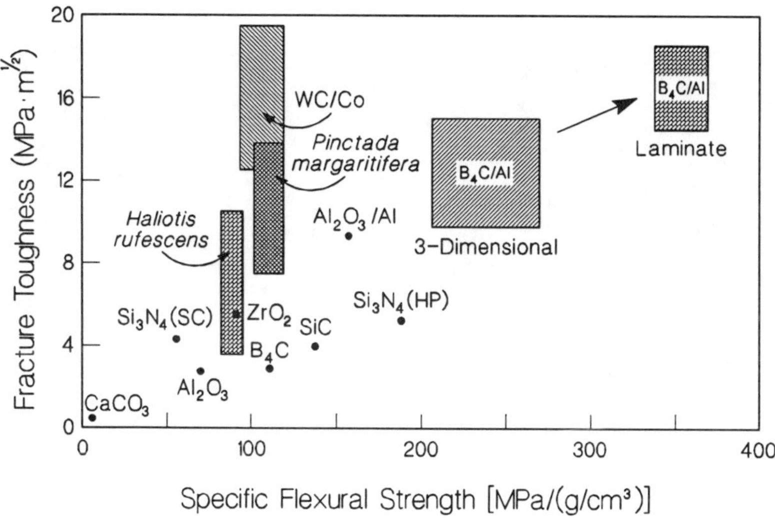

Figure 5. Mechanical properties of nacre of abalone and pinctada (pearl oyster) compared to some major ceramic and cermet materials.

The investigation on crack propagation behavior in nacre[84,85] reveals that there is a high degree of tortuosity not seen in the more traditional brittle ceramics, such as Al_2O_3 or in high toughness ceramics, such as in fiber-reinforced SiC[97] and particulate-reinforced ZrO_2[98] composites. The surfaces of fractured samples indicate that a major crack has meandered around the $CaCO_3$ layers resulting in a rough fractured surface, similar to that seen in fiber-reinforced ceramic composites where a pull-out mechanism operates.[99] Examination of micrographs recorded at higher magnifications indicates either sliding of the $CaCO_3$ layers upon organic matrix, suggesting the resolved stresses are lateral to the layers, or formation of organic ligaments between the layers, when the stresses are normal to the layers (Figure 6). The latter case, stretching, indicates that the interface between the organic and the inorganic phases is strong and that the organic phase acts as a strong binder.[84,85] Sliding, or bridging, of the macromolecular assemblages that make up the composite organic matrix under different modes of stresses is an excellent example of the unique quality of biological materials in terms of their structures and resulting properties. Several toughening mechanisms, therefore, may be proposed:[100] (i) crack blunting/ branching, (ii) microcrack formation, (iii) plate pull-out, (iv) crack bridging (ligament formation), and (v) sliding of $CaCO_3$ layers.

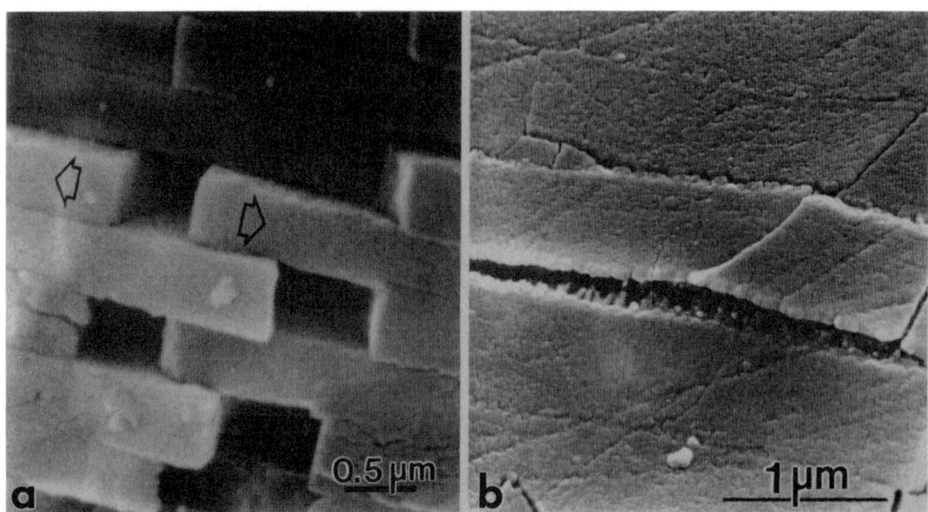

Figure 6. SEM images recorded near an indentation performed on an edge-on configuration in nacre of abalone. Either sliding of aragonite platelets (a) or ligament formation by the organic phase (b) takes place depending on the resolved applied stress (longitudinal and transverse, respectively). Both mechanisms are responsible for the high toughness of the nacre.

The high degree of tortuosity seen in crack propagation[85] may be due mainly to crack blunting and branching.[97,99] Tortuosity alone, however, is not a major toughening mechanism in these composites, because it cannot account for the many orders of magnitude increase in toughness. The major toughening mechanisms, therefore, are sliding and ligament formation.[85] Similarities exist between the deformation of a nacreous seashell[30,85] and a metal[101] in the sliding mechanism in terms of forming deformation bands in the bulk that appear as striations on the surface. It is likely that this complex deformation mode may be the main mechanism of energy absorption during the propagation of cracks, which needs to be further investigated quantitatively in terms of its energetics.

The strength of nacre, on the other hand, may be related to several factors, including the size and structure of the aragonite platelets and the interfaces between the inorganic and the organic components. From the limited thickness of the largest flaw (i.e., the thickness of the platelet) 0.5 µm, the increase in the theoretical fracture strength[102] of aragonite would be about 200 MPa, comparable to the measured value of 185-220 MPa in our studies.[85] Therefore, the rule-of-mixtures[103] may account, at least, for part of the value of the measured strength.[83-85,100]

3.0 DESIGN GUIDELINES FOR PROCESSING BIOMIMETIC LAMINATED COMPOSITES

The lamination microarchitecture of present day impact resistance materials originates from ancient armor.[104] Biological structures, such as those in insects,[32] egg shells,[105] and sea shells[91-94] also have similar architectures, but at much smaller dimensional scale. In fact, the best armor has a double-architectural design in which the front section is hard, for the effective stopping of the projectile, and the back section is soft (but strong) for the absorption and dissipation of the kinetic energy from the projectile.[104] This is basically analogous to the shell of abalone. In cross-section, the hard front consists of long calcite crystals perpendicular to the shell plane (and in the direction of the projectile, ready to confronting it), and the back is tough nacre (laminated aragonite and organic matrix) as schematically illustrated in Figure 7 (also see Figure 10). In a ceramic/ceramic composite design, on the other hand, even in laminated architecture, weak interfaces are a necessary condition for increased toughness, with a sacrifice in the expected deterioration of the overall strength of the composite.[106,107] Contrary to the accepted materials design criteria in the role of interfacial strength in synthetic materials (as described in the previous section) in the structural design of biological materials such opposing effects are circumvented by the role of macromolecules due to their composite structures and resulting multifunctionality. Hence, there seems to be no need for a sacrifice in any of the properties, and therefore, both the toughness and the strength of the biocomposite increase, as seen in the laminated structure of nacre.

The unusual mechanical properties of nacre, absent in synthetic ceramics and composites, may, therefore, derive from: (i) the intrinsic properties of the constituent phases (brittle inorganic phase and soft organic matrix); (ii) the highly ordered organization of ceramic and biopolymer layers and their detailed structures, including the interface structures and properties; and (iii) the size of the ceramic and biopolymer layers which may be critical factors due to the intrinsic properties of biogenic aragonite and the organic matrix. Despite the fact that many of the structural features of abalone remain

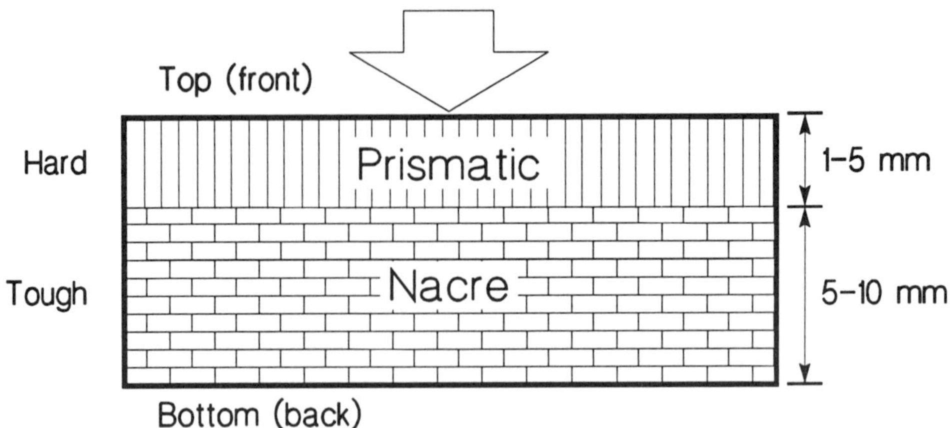

Figure 7. Schematic illustration of the cross-section of the abalone or pinctada shell, an ideal armor material with a hard front and a tough back.

a mystery (especially of the soft tissue), we can distinguish some guidelines using the knowledge of their excellent mechanical properties and structural architecture at the micro- and submicron levels to aid us in the processing of synthetic microstructures through biomimicking. Based on the previous studies of nacre,[30,85] the design of high-strength/high-toughness synthetic composites should incorporate the following:

(i) a laminate thickness of the hard and brittle component of less than 1 μm and a soft component of less than 100 nm, with an approximate ratio between 5:1 and 10:1

(ii) a highly plastic soft phase (deformability > 100 %)

(iii) strong interfaces between the soft and hard phases (so that the interface does not fail during crack propagation)

(iv) a soft phase able to bind to the surfaces of the hard component (strong interfacial bonding) and provide plasticity to the overall composite structure (for sliding and metal-like deformation behavior) or form ligaments to constrain crack opening (through crack bridging), depending on the state of resolved applied stress.

In actual processing of laminated composites, these guidelines may be difficult to apply. For example, there is currently no practical way to produce structural laminates with layer thicknesses thinner than about 10 μm[108,109] (except in *in-situ* laminates formed through phase transformations in which overall architectural design is somewhat limited due to thermodynamics or kinetics limitations).[8] It is also difficult, if not impossible, to control the thickness ratio of the hard and soft components at these small dimensions. Nevertheless, the guidelines stated above, remain useful as they present the ultimate structural features. In the design of synthetic laminates, high hardness ceramics such as BN, B_4C, TiC, ZrO_2, and Al_2O_3 can be used as the brittle component. Highly plastic (superplastic) metals, such as Al and Cu (and their alloys), or organic polymers, such as polyethylene and polypropylene, may be good candidates for the soft phases. These constituent phases which give the best combination of properties, with emphasis on strong interfaces, could be used if they could be processed.

Some of the property requirements stated above have been met to a certain degree in the B_4C-Al[100,109] and B_4C-polymer[108] laminated systems designed for use as impact resistant materials. In one synthesis strategy for processing of B_4C-Al composites, porous B_4C layers with thicknesses below 100 μm, and as low as 15 μm, were tape-cast between thin sheets of Al. The stacks were heated to induce infiltration and to allow bonding between Al and B_4C without an excessive reaction. The resulting composite displays a structure in which both the Al-infiltrated B_4C and the alternating layers of pure Al form continuous films.[100,109] The overall architecture provides alternating layers of hard and soft components and strong interfaces. As a result, fracture toughness and fracture strength both show a 30 to 40 % increase over the isotropic B_4C-Al composite with the same phase ratios (Figure 6) in which Al and B_4C have a three-dimensional interpenetrating network.[100,109] The mechanical properties of B_4C-Al laminates, in terms of K_{IC} and specific fracture strength, are currently the best that can be achieved among the present cermet systems.

Despite the fact that improvements have been achieved in the mechanical properties of laminated composites based on biomimetic architecture, as in the example of B_4C-Al, these improvements have not yet come close to the superior properties of nacre over monolithic $CaCO_3$. This may be due primarily to insufficiently thin laminate layers. Thicknesses below 1 μm in the inorganic layers and below 100 nm in the organic layers (soft phase) are needed. Second, as will be clear from the discussion in the following section, both the inorganic and organic layers have complex structures, in terms of their crystallography, substructures, and morphology.

4.0 DETAILED STRUCTURE OF NACRE

In a layered composite design,[110] the most significant structural features are: (i) properties of the hard and soft components, (ii) interface structures and properties, (iii) thicknesses of the laminates and their ratio, and (iv) geometry of the lamellae. In the case of nacre, although many structural factors are known, some of the most significant questions are still unanswered. In nacre, for example, it is possible that the aragonite lattice may incorporate organic molecules.[90] If this is so, then the bulk properties (e.g., moduli) of the biogenic aragonite will differ significantly from those of geological aragonite.[30] As shown in Figure 8, biogenic aragonite contains a significant density of dislocations (in addition to other substructural features),[111] contrary to geological aragonite (an ionic crystal) in which dislocations could not be observed.[112] In nacre, the lamellae do not form simple continuous layers; instead the inorganic component, in the form of platelets (bricks), interrupts the continuity of a given layer with each platelet surrounded by a thin film of organic matrix. Even less is known about the composition and molecular and conformational structure of the organic matrix.[86,89,90,113-118]

Among all the features of nacre, the interface structure and properties are the least understood. Interfaces not only allow sliding of aragonite platelets and the formation of organic ligament under stress (both necessary conditions for the excellent properties), but they are also the sites of the nucleation and growth of the aragonite layers.[113-115,118,119] For these reasons, structural correlation of aragonite and the organic matrix deserve closer examination.

In the following sections we summarize the current understanding of morphological and crystallographic relationships among the structural components of nacre. First, we present an overview of these relationships between the inorganic building blocks. From this, we construct a model for the possible molecular conformation of the underlying organic matrix. We find that the model not only allows the crystallographic but also the morphological relationships among building units in nacre and forms a basis for shape formation and growth of the shell.

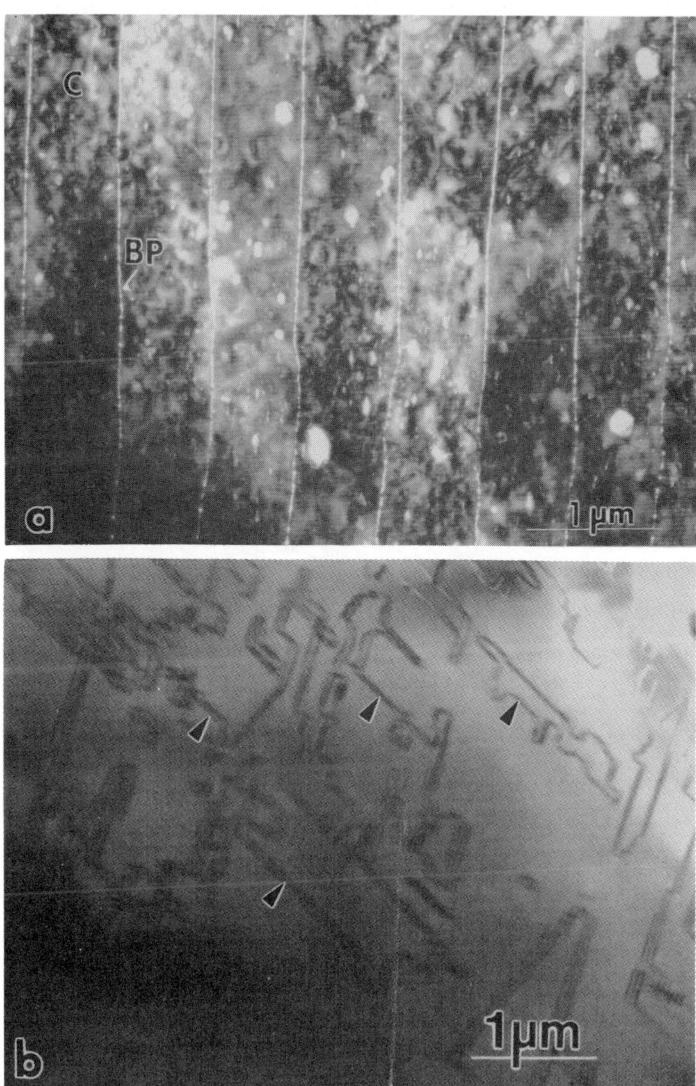

Figure 8. (a) Platelets of biological aragonite in an edge-on view of nacre of pinctada, and (b) geological aragonite (sample from Aragon, Spain) display substructural details. Note high dislocation density (in this orientation) in (a) but only antiphase boundaries in (b). [Twins in (a) are in face-on configuration and, hence, are not seen.]

4.1 STRUCTURE OF NACRE

Red abalone (*Haliotis rufescens*) belongs to the mollusk species in the family of gastropods. Gastropods first evolved about 500 million years ago, and, with little change, they have survived to the present time (abalone is, thus, considered a "living fossil").[21,91-94] The shell of the abalone is ear-shaped, with a large opening and a small spire (Figure 9). The shell has respiratory pores (about 20 in the adult) but only the last 4 to 5 are open at a given age. The organism has a large foot which, in the juvenile period, grabs onto a rock, allowing adult mollusk to forage for algae and plants on the bottom. Although other abalone species are much smaller, the shell of *H. rufescens* grows by about 2.5 cm per year and can reach 30 cm in diameter after which it only thickens (becoming more than a cm).

A longitudinal cross-section of the red abalone shell displays two types of microstructures: an outer prismatic layer and inner nacreous layer (see Figures 7, 9, and 10). Two forms of $CaCO_3$, calcite (rhombohedral, $R\overline{3}m$) and aragonite (orthorhombic, Pmnc), constitute the inorganic phase of the composite in the prismatic and nacreous layers, respectively. These architectures, especially the highly ordered inorganic component (aragonite platelets), are possible to study crystallographically with the aid of electron microscopy directly and in great detail. This is mainly because electron radiation damage poses less of a threat during the observation of the $CaCO_3$ crystals than does the organic matrix.

The structures of shells are seen in the SEM images in Figure 10, from the fractured surfaces of red-abalone and nautilus. The images display stacked platelets (aragonite) in nacre on the inner portion, and elongated crystallites (calcite) in the prismatic section on the outer portion. Platelets in nacre are typically arranged in either columns (abalone and nautilus) or sheets (pinctada). The stacking is not random and, in a fully grown specimen, resembles a brick and mortar architecture. The platelets vary in thickness; an average of 0.25, 0.4, and 0.5 μm in red abalone, nautilus, and pinctada, respectively, and edge length is 5-10 mm. Organic matrix is between 10 and 50 nm thick. The dimensions of the aragonite platelets and the organic matrix vary depending on the site of the shell from which the sample is extracted and the species of the mollusks.

Figure 9. Schematic drawing of (a) a top view and (b) an inside view of *Haliotis rufescens.*

4.2 STRUCTURE OF THE ORGANIC MATRIX

As described by other researchers,[90,119,120] the organic matrix is thought to be a composite of macromolecules stacked in a sandwich form. This is primarily based on indirect investigation of the characteristics of the biological macromolecules that are extracted from the shell through decalcification.[90,114,119,120] Macromolecules are divided into two groups: those soluble in weak acids, and those that are not.[90] In the former case, the proteins extracted were found to contain aspardic and glutamic acids (see references in Table-II). These amino acids are known to be major components of structural proteins that form sheet-like structures, such as in β-pleated sheets, that are found throughout the animal kingdom.[23] An extensive literature survey reveals the major amino acid compositions of the proteins in various species of mollusks given in Table-II. There is also speculation of the composition of the insoluble portion of the organic matrix, but it is probably much less accurate. It is possible, based on many investigations, that this portion of the matrix may constitute polysaccharides in addition to proteins (see Table-II).

Figure 10. Images of fractured surfaces of abalone (a) and nautilus (b) exhibit both the prismatic (p) and the nacre (n) sections.

The major flaw in each of these studies has been that the macromolecules extracted from the shell have not been specific to a certain section in the shell. In fact, the shell is often pulverized, with both the calcitic (prismatic) and aragonitic (nacre) sections intermixed before decalcification. The organic macromolecules are expected to have different stoichiometric compositions, and hence different conformations with respect to each other. These differences may, in turn, greatly affect the structure, crystallography, and the geometry of the inorganic crystals. The analysis which gives a mixture of all the biological macromolecules in a shell, therefore, has so far not been adequate in describing the true composition and the structure of the macromolecules that make up the nacre structure. (Several investigators - D. E. Morse,[121i] University of California, Santa Barbara, D. L. Kaplan,[121ii] Natick Research Center, ARO Lab, and C. E. Furlong, the University of Washington,[121iii] are working towards isolating macromolecules from different portions of the shell.)

Nevertheless, based on the findings so far, and limited direct analysis, the organic matrix may actually have a sandwich structure containing three organic sublayers, each with its own unique composition and molecular conformation.[114] According to this scheme, the central portion of the organic matrix is composed of chitin (polysaccharides), which is surrounded on both sides by β-pleated protein sheets. The layer next to the inorganic is then composed of acidic proteins. In this scheme, while the chitin provides the structural (mechanical) stability to the composite forming a backbone (framework macromolecules), the β-pleated sheets act as the substrate and provide the biomineralization sites (nucleator macromolecules). The acidic proteins are there to fill in the gaps between the organic matrix and the aragonite crystals. Neither the composition of these layers nor their structure has yet been clearly identified. The investigation of the structural relationship between the organic and the inorganic layers is, therefore, far from complete.[30]

Similarly, it may be possible to directly study the organic macromolecules in the TEM if proper molecular markers are developed.[122,123] For this, isolation of each of the major macromolecules from the organic matrix seems to be necessary. Once this is done development of antibodies and proper stains will follow. These stains can then be used in ultramicrotomed thin sections to identify locations and concentrations of each of the macromolecules. A preliminary work has been carried out to reveal the structures of the macromolecules in samples that are either ultramicrotomed or low temperature ion-beam milled.[124] Two such micrographs are displayed in Figure 11. Figure 11(a) reveals layered structure of the organic matrix in an ultramicrotomed section.[124]

Table-II. Amino acid compositions of the mollusk shells used in this study (table is made using the literature data)

		Asx	Thr	Ser	Glx	Pro	Gly	Ala	Val	Tyr	ref.
H. rufescens	W	25.0	?	10.2	8.2	?	48.5	4.52	?	?	i
	N	20.0	2.0	9.2	4.3	3.6	18.4	17.1	?	?	ii
	P	20.0	11.0	6.8	5.7	5.7	11.0	6.8	?	?	ii
N. pompilius	WI	7.1	1.3	9.8	4.5	0.5	35.3	25.0	1.4	0.6	iii/iv
	N	2.0	1.3	6.3	6.7	0.0	19.7	48.0	2.3	0.0	v
	NS	26.1	4.8	7.9	6.6	4.6	23.6	4.4	1.5	6.4	vi
P. margaritifera	P	9.4	2.9	3.9	2.2	5.6	22.3	3.2	?	5.5	vii

N: Nacre only; P: prismatic only; NS: nacre-soluble proteins; W: whole shell; WI: whole shell, insoluble only.

i. M. Cariolu and D. E. Morse, "Purification and Characterization of Calcium-binding Conchiolin Shell Peptides from the Mollusc, *Haliotis rufescens*, as a Function of Development," J. Comp. Biol. B, **157**, 717-729 (1988).
ii. N. Nakahara, G. Bevelander, and M. Kakei, "Electron Microscopic and Amino Acid Studies of the Outer and Inner Shell Layers of Haliotis rufescens," Venus, **41** [1] 34-46 (1982).
iii. M. F. Voss-Foucart, "Assais de solubilization et de fractionationnement d'une concioline (nacre mulare de *Nautilus pompelius*, mollusque cephalepode)," Comp. Biochem., **26**, 877-886 (1968).
iv. E. T. Degens, D. W. Spencer, and R. H. Parker, "Pleobiochemistry of Molluscan Shell Proteins," Comp. Biochem. Physiol., **20**, 553-579 (1967).
v. S. Weiner and L. Hood, "Soluble Protein of the Organic Matrix of Mollusk Shells: A Potential Template for Shell Formation," Science, **190**, 987-989 (1975).
vi. G. Goffinet and C. Jeuniaux, "Composition chimique de la function "nacrpine" de la conchioline de nacre de *Nautilus pompelius* Lamarck," Comp. Biochem. Physiol., **29**, 277-282 (1969).
vii. S. Tanaka, H. Hatano, and O. Itasaka, Biochemical Studies on Pearl. IX. Amino Acid Composition of Conchiolin in Pearl and Shell," Bull. Chem. Soc. Japan, 33, 543-545 (1960).
viii. H. Nakahara, M. Kakei, and G. Bevelander, "Fine Structure and Amino Acid Composition of the Organic Envelope in the Prismatic Layer of Some Bivalve Shells," Venus, **39** [3] 167-177 (1980).

The contrast difference is due to differential staining of the organic sub-layers by uranyl acetate due to their differences in composition and structure. A detailed structure of the interface between organic layer and aragonite is shown in Figure 11 (b). The image was taken from a sample that was low-temperature ion-milled and then slightly etched by gold citrate.[124] The preferential etching of the aragonite platelets gives them a saw-tooth appearance. The light acid used that causes this feature also dissolves away the acid soluble portion of the organic sublayers. As a result, central sublayers of the organic matrix are exposed. This is the part we think is made up of polysaccharides since they are known to be resistant to such treatment. Further studies are underway to correlate these characteristic features related to the structures of aragonite and the organic matrix.

From the point of biomimetics it is essential to understand what the function of the organic matrix in nacre is in controlling nucleation of the inorganic crystals, their shape formation and growth. In future crystal engineering, in making nanostructures and laminated composites based on biological hierarchical composites using either synthetic or biological macromolecules, the first requirement is to understand the mechanism of inorganic-organic interactions in biological systems. Further detailed investigation, especially in a hard tissue like nacre having a relatively simple structure, appears to be well warranted. The following section attempts to address this issue.

4.3 CRYSTALLOGRAPHY OF ARAGONITE PLATELETS

Prior studies have postulated the crystallinity of the organic matrix that it contains chitin and silk fibroin-like proteins as the outer and middle sublayers of a sandwich structure and that these macromolecules are known to self-assemble in crystalline arrangements.[90,114-115,119,120] The outermost sublayer in the organic matrix, the one in contact with $CaCO_3$, consists mostly of soluble charged macromolecules (proteins) and presumably acts as the binder to the inorganic phase. This scheme suggests that active sites on the matrix align with those on the $CaCO_3$, i.e., either Ca^{2+} ions[90,114] or CO_3^- sites (ionotropy mechanism).[113,119]

It has been impossible to study both the organic and inorganic crystals simultaneously, and the structural relationships between the components of the nacre exist only as a conjecture.[90,113-115,119,120,125-128] Bulk studies, performed by X-ray diffraction, on the nacre revealed that the aragonite platelets are organized with their [001] axis perpendicular to the layers.[90,114,128] It has been postulated that the axes within the layer plane in each platelet are oriented randomly.[90,114] Furthermore, it was assumed from this scheme that each aragonite platelet grew on the crystallographically related organic template, which itself has a local random orientation.[90] From the composition of the insoluble fraction of the organic matrix, i.e., the inner crystalline sublayers which contain a high fraction of aspartic and glutamic acid, it might be possible to deduce a self-

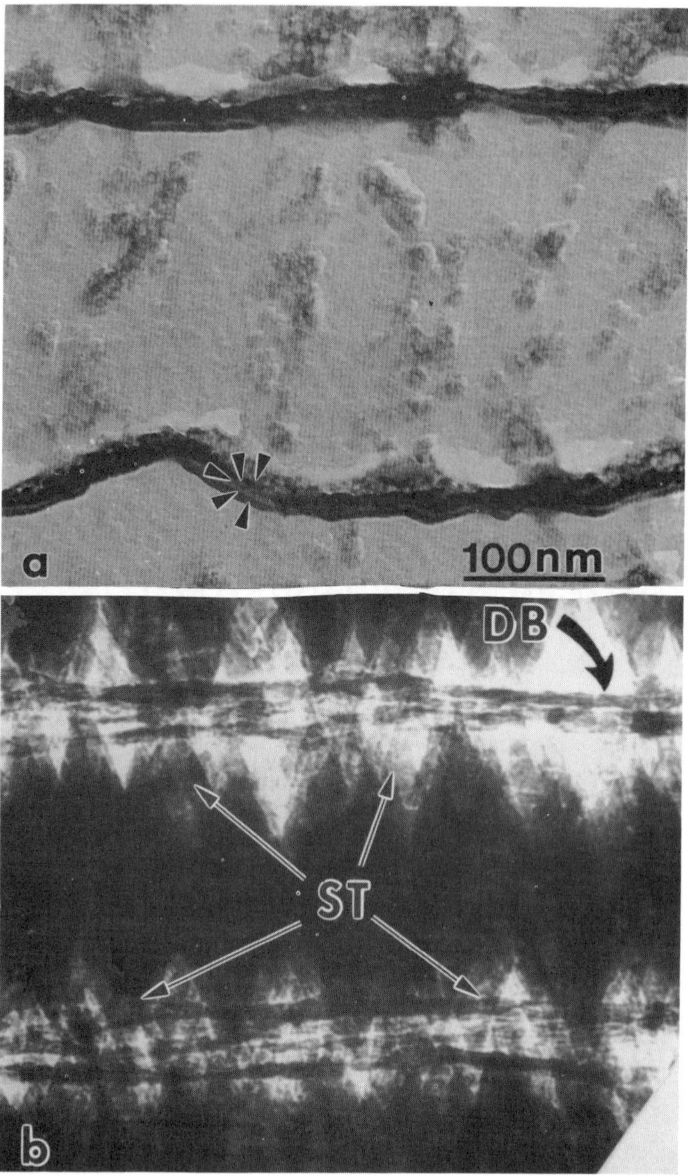

Figure 11. Structure of the organic matrix in nacre of abalone. The sample in (a) was ultramicrotomed, decalcified, fixed, and stained. Five sublayers (arrows) are exposed in the organic matrix, possibly due to their differential staining. The image in (b) was taken from an ion-beam milled sample that was slightly acid etched. It reveals insoluble (central) portion of the organic layer DB (dark bands) and the saw-tooth (ST) appearance of the etched aragonite/matrix interface.

assembled structure that would be related commensurately to the aragonite lattice along the [001] projection.[90,114,120,128]

In the present work, each aragonite crystallite was analyzed by electron diffraction separately and its crystallographic orientation relationship was established with respect to its neighbors, both on the same layer and across the thickness of the nacre, hence, enabling us to complete a three-dimensional picture.[30] First, we found that adjacent platelets on the same layer belong to the same [001] zone axis with a slight rotation among them. The question whether there is any crystallographic relationship between the a and b axes of platelets on the same layer was also answered; we found that they are twin-related with a twin plane of {110} type.[30] In this scheme, all of the platelets on the same layer are twin-related whether they share a boundary or not, constituting *first generation twins* since this twinning takes place at the highest spatial scale [Figure 12 (a-b)]. Further analysis indicated that each platelet consists of several domains which are crystallographically related (Figure 12 (c-f)). The diffraction patterns from all the domain boundaries show twin reflections with domains belonging to either (110) or (110) variant. Adjacent domains, therefore, are twin related and constitute the *second generation twins.*

The angle between each pair of domains in an ideal hexagonally-shaped platelet with six twin-related domains would be 60°. This is not possible, however, since the outer edges of the platelets are parallel to {110} planes. The angle between each pair of planes - for example, between (110) and (110) - is 63.5°; this leaves a 3.5° unaccounted for. This induces strain into the aragonite lattice and must be accommodated by some structural deformation, such as, by slip or twin formation. In the present case, nanometer-scale twins form on {110} planes, shown in Figure 12 (g), that are similar to the growth twins in geological minerals. Since they occur at the smallest scale, we refer to them as the *third generation twins.*[30] (A portion of the lattice stress created by the 3.5°-strain can also be accommodated by the misalignment of adjacent domains, as frequently observed. This misalignment, however, cannot account for all the strain accommodation, as the interfaces between the domains show a high degree of coherency.) These three twin structures cover a size scale of six orders of magnitude, from the nanometer to the submillimeter, and reveal, for the first time, a hierarchical structure in a biological hard tissue.

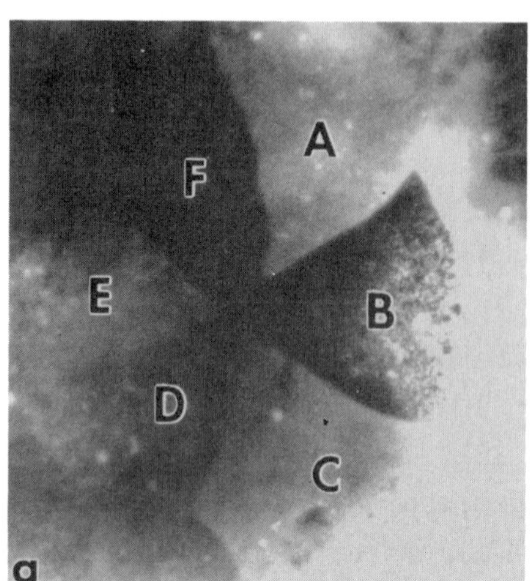

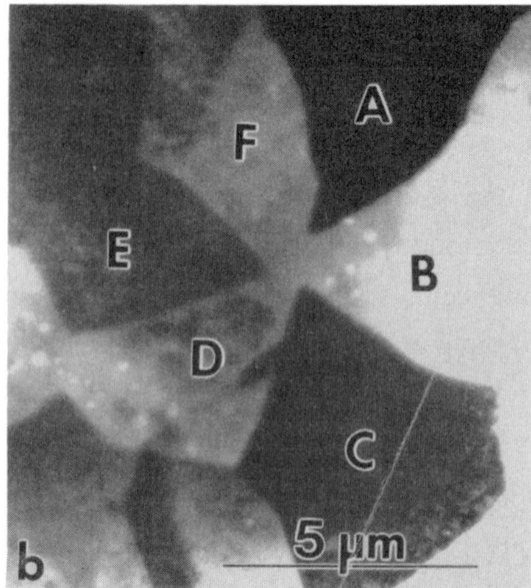

Figure 12 (a-g). TEM images reveal crystallographic relation between aragonite platelets and domains. Images in (a) and (b) reveal that A-C-E are twin related to B-D-F, as confirmed by electron diffraction (image recorded with a slight tilt about the [001] axis that is normal to the plane of the paper).

4.4 THE CONFORMATION OF ORGANIC MACROMOLECULES

The interaction between the species on an organic substrate and a crystalline organic phase must include geometrical, electrostatic, and stereochemical interactions.[113-115,119,120,127-131] Therefore, the nucleation and growth of the crystals will be influenced by both the nearest neighbor interactions and those higher on the scale. The aragonite crystal structure belongs to the space group Pmcn (No. 62) with lattice parameters a = 4.94 Å, b = 7.94 Å, and c = 5.72 Å.[132] The location of Ca^{2+} ions in [001] projection is shown in Figure 12(a) giving the crystal a pseudo-hexagonal symmetry. In the unit cell, CO_3^- groups would reduce the symmetry to an orthorhombic form.[132] This is an important physical characteristic in terms of the crystallographic relationships of the hierarchical twinned components of the nacre aragonite and the stereochemical relationship that might exist between the aragonite and the macromolecules in the organic matrix.

Figure 12 (cont.). Images in (c) and (d) are from a platelet in (e) containing four domains. [001] diffraction pattern (f) taken from a domain boundary in (e) reveals that the domains are twin related.

The nucleation and growth of the aragonite crystals may involve both Ca^{2+} and other ions, but for simplicity, only the arrangement of Ca^{2+} ions is illustrated in Figure 13. True hexagonal closed-packed formation would require an axial ratio of b/a = 1.63; however, the observed value is only 1.69. In the actual aragonite lattice, the Ca^{2+} ions are not in contact but are separated by O^{2-} ions, and the pseudo-hexagonal arrangement refers to the centers of Ca^{2+} ions rather than their actual packing. Since the same {110} twinning takes place at all length scales, superimposition of the lattices on all three possible twins with a 63.5° angle with respect to each other generates a new superlattice structure based on the coincidence lattice sites (CLS)[133] and is called *superstructure*[30,134] (Figure 13(c)). Taking the actual distance between the Ca^{2+} ions as 3.94 Å, the calculated distance between these lattice points would be about 30-40 Å. If the nucleation and growth of the aragonite platelets takes place on the underlying organic matrix, then the geometric configuration of the active sites for binding Ca^{2+} ions on the organic matrix must accommodate this superlattice and thus all twins in the nacre. To explain this phenomenon, we assume that the binding sites on the template form a single crystalline pseudo-hexagonal lattice, or integer multiples of it. Many two-dimensional membranes tend to form hexagonal lattices during self-assembly.[135] This hypothesis (that the organic matrix may be a single crystal) is an essential structural feature for the formation of the highly organized platelets in nacre.[30,134] The local crystalline organization of the matrix proposed earlier,[90,114] with no relationship between the neighboring areas and therefore no long-range order, would result in aragonite crystals with no definite crystallographic orientation relationships.

By tracing the possible twin boundaries, the superlattice allows the generation of the hierarchical twin structure and all the shapes, geometry, and crystallography-related features, including five-edged platelets with 90°-domains, sixfold symmetry of platelets, and six- and three-edged domains.[30,134] The fact that one can generate all the possible configurations in this way illustrates again that a pseudo-hexagonal template structure (and not the lattice of a single domain) might be a possible solution for the structure of an organic matrix that can accommodate all the twin relationships. The ultrastructure in the organic matrix would be single crystalline not only on the flat surface but also through the transverse direction of nacre as demonstrated from the twin orientation-relation of the crystals across the thickness direction as well.[30] The platelets do not leave space when packed on a given layer because of the crystallographic and geometrical requirements, hence called *space filling tiles*. The tiles (aragonite platelets) can have several different edges (3-, 4-, 5-, and 6-edged) but still have regular shapes. Based on a mathematical description,[136] we call them *multiple tiles*.[137] Since the tiles also have domains at lower dimensional scale, we can call the configuration of the tiles within the nacre in three-dimensional space as multiple tiling with hierarchical twins.[137]

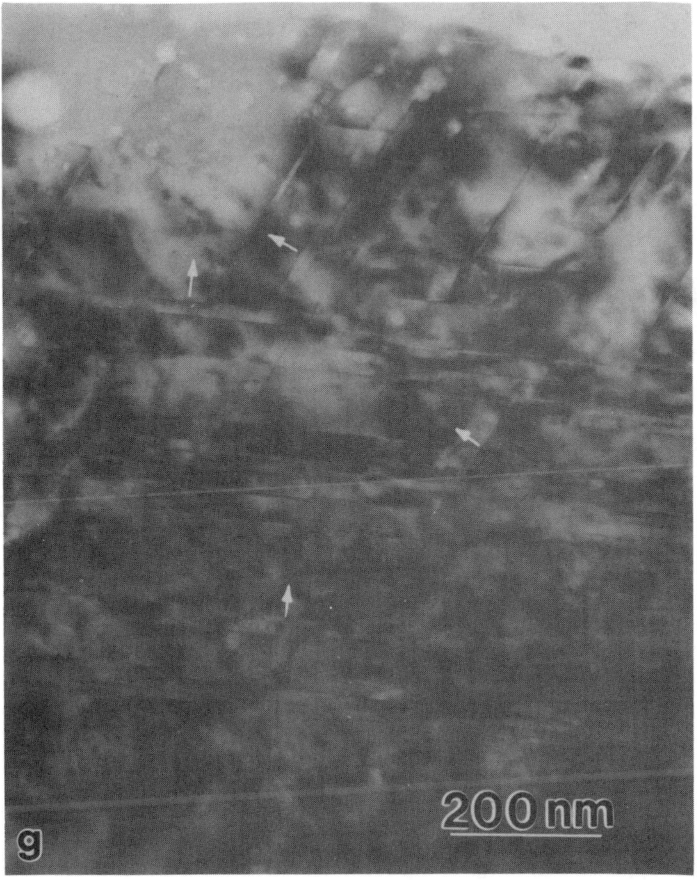

Figure 12 (cont.). (g) Bright Field image taken in [001] orientation shows the ultrafine twins on two {110} variants within a domain.

5.0 GROWTH AND SHAPE FORMATION OF THE SHELL: MORPHOGENESIS

Although six-fold twin structures also occur in geological aragonite,[111] the hierarchical arrangement in nacre is unique in that each platelet is separated from the others during the early stage of growth.[134,137] This is illustrated in Figure 14 which shows the layers of aragonite are separated on the growing edge of the shell by a thin film of organic matrix, and the new platelets grow on it independently. In geological aragonite, by contrast, the mimetic twin domains grow in contact, one after the other, and each is influenced by the presence of another.[111,112] In nacre, however, even after crystalliza-

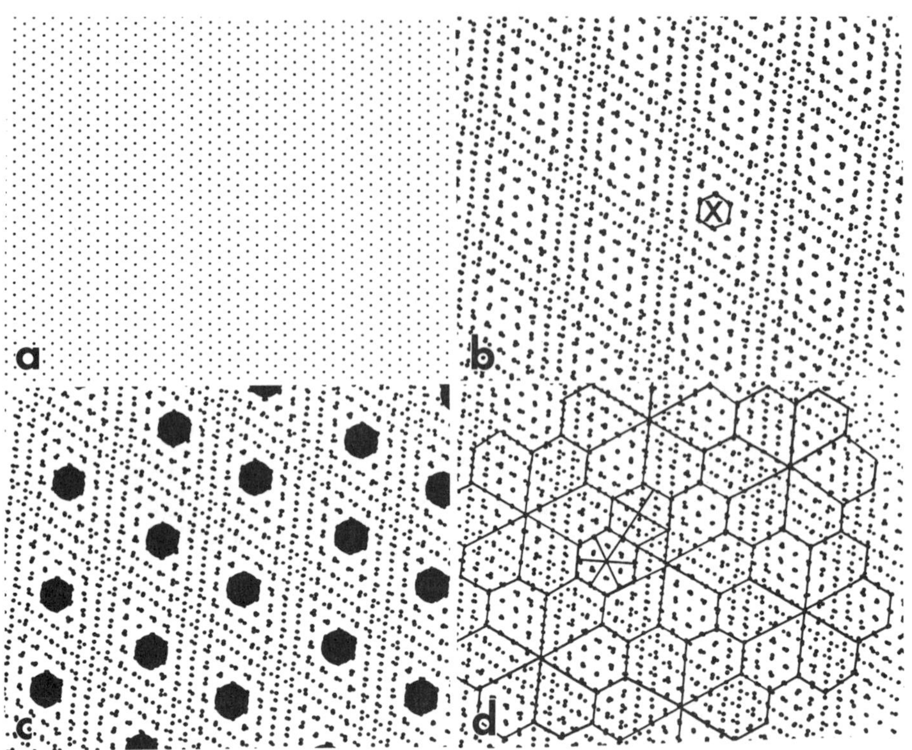

Figure 13. (a) The model of the aragonite lattice in [001] projection with only Ca^{2+} ions highlighted. (b) Superimposition of three lattices each with a successive rotation of 63.5° (angle between {110} planes) produces a Morié fringe pattern. (c) Highlighted coincidence lattice sites produce a pseudo-hexagonal pattern. (d) Platelets, which retain the crystallographic and morphological relations, can be drawn based on the CLS model.

tion is complete, the platelets remain separated from each other by an organic membrane.[132] The fact that separate platelets grow simultaneously and, yet, have a definite crystallographic orientation relationship suggests that the growth process might be mediated by the organic template beneath each of the individual crystals, as proposed earlier.[114,115,119,120]

As can be determined from Figure 14, which was recorded from the growth edge of a juvenile abalone, the organic membranes actually form layers of sheet with an empty space between them which would eventually be filled with the inorganic material. Each

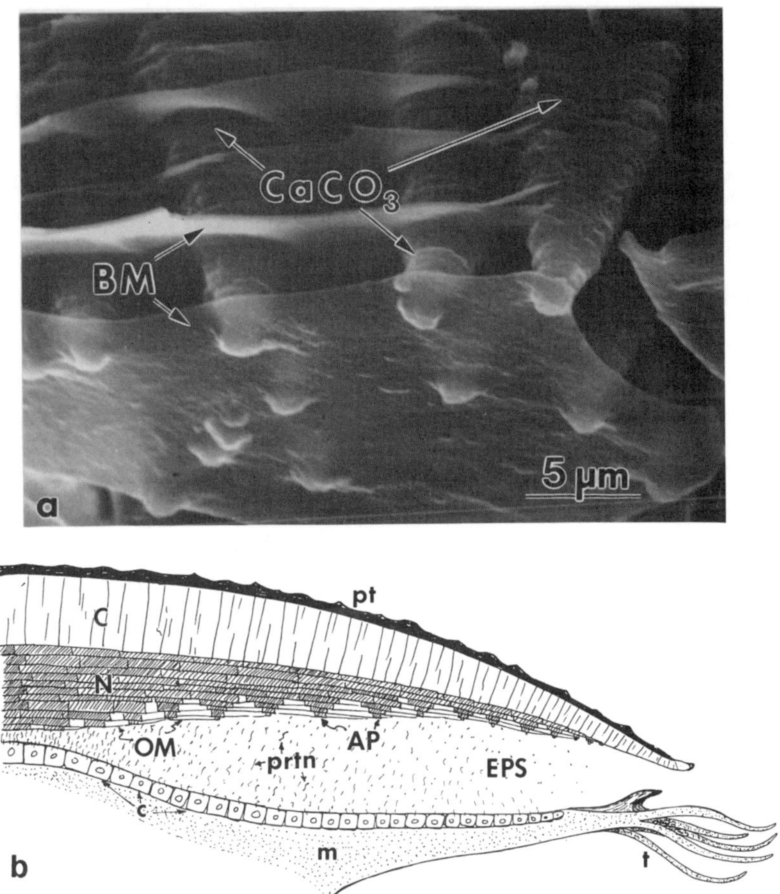

Figure 14. (a) Secondary electron image of the growing edge of a juvenile red abalone in an edge-on view shows organic matrix (BM) between the newly forming aragonite platelets ($CaCO_3$) in a columnar organization. Notice the bottommost organic layer with newly formed aragonite nuclei. (b) A schematic drawing illustrates the growing edge of the abalone shell. C: prismatic layer; N: nacreous layer; EPL: extrapallial space; pt: periostracum; c: cells; prtn: self assembling proteins and polysaccharides; m: muscle of the foot; t: tentacles, OP: organic matrix; and AP: aragonite platelets.

layer occasionally separates into several sublayers, and this is repeated for many layers, seemingly in a random fashion, throughout the thickness. This suggests a hierarchical organization of the organic membrane as well. The production of further organic layers originating from a single membrane also has significant implications in terms of the mechanism of self-assembly of the macromolecules that make up the framework of the organic matrix. Although attempts have been made,[119,126,135-137] a clear description explaining how the successive single layers form in the nacre, or any other hard tissue has not been established. The best description so far has been the presence of compartments which would initially form as small boxes on top of the older and much larger ones. These would then eventually grow as the inorganic crystallites form within them. In this "compartmentalization" scheme, it is not clear how the layers actually come into being in the first place and how each compartment would form and be enlarged.

In contrast, we show that successive layers form from an original membrane as subsidiary layers, which would themselves act as originators of further subsidiary layers in a cascading fashion.[137] This mechanism is similar to the reproduction of cells in organisms in making new cells and subcellular features (such as organelles). If this scheme is correct, then the layers must be in closer scrutiny by the organism than what has been hitherto thought as merely being formed through self-assembly. In other words, the cell membrane formation is a further step up in the hierarchy of making cells than the self-assembly of macromolecules which is a quaternary step in the organization of macromolecules in organisms, such as protein clusters, or structural polysaccharides. It can also be noted, in this case, that both proteins and polysaccharides are thought to be present in the organic matrix structures, adding more evidence that the organic matrix might be a membrane in the sense of physiological terms. A membrane therefore, having proper proteins in their respective locations, both within the composite layers of the organic matrix and on certain locations of the membrane surfaces, would serve the function of being structural proteins, and of transport proteins regulating the passage of the inorganic ions in and out of the membrane for the formation of $CaCO_3$ on the surface. This description on the formation and the function of the organic matrix as, what we will call, a *pseudomembrane*,[137] is conjectural and requires further study. These studies would be essential to find the true nature of structures and growth of hard tissues.

The shapes of many mollusk shells can be described mathematically by a helicospiral.[138,139] This is demonstrated schematically in Figure 15(a) for abalone shell. In two-dimensions (x-y plane) the equation of the spiral is: $r = ae^{\phi}$ where r is the radius of the spiral at any position of the angle ϕ, a is a constant specific to a species of the mollusk. The depth is given to the shell by a variation in the third dimension, z. What

Figure 15. (a) Logarithmic spiral. (b) A schematic illustration of growth spirals of abalone shell. (c) Local spirals of aragonitic platelets in nacre (secondary electron image).

has not been recognized so far, neither by biologists nor by mathematicians, however, is the possibility of the design of the overall shape of the shell based on the underlying organic matrix within the hard tissue.[138,139] This design may be a result of the structural interrelationship between the crystallography of the aragonite lattice and the crystallographic conformation of the Ca^{2+}-binding (or CO_3^- -binding) matrix nucleator proteins. For instance, assuming that macromolecular conformation based on the proposed superstructural model above provides a template for the $CaCO_3$ formation, then the growth of an individual aragonite platelet follows a spiral [Figure 15(b)]. This growth scheme may allow adjacent platelets to grow in a spiral manner, as demonstrated in the SEM image in Figure 15(c). Spiral growth of platelets in local regions along and behind the growing edge of the shell would lead to the growth of the shell at hierarchically higher scales, eventually leading to the final specific shape of each of the mollusk species. In this proposed growth scheme, the overall morphology of the shell is designed at the molecular scale by the organic matrix.[137] In terms of biomimetics, this hypothesis, if proven correct, may have implications in net-shape formation of technological ceramics, and, therefore, constitutes another important area of future investigation.

6.0 SUMMARY AND FUTURE DIRECTIONS

We have reviewed the results of some of the recent studies on mechanical properties and structure of the nacre section of mollusks. Nacre, which is about 95 vol. % aragonite and about 5 vol. % organic macromolecules, has fracture toughness and fracture strength properties that are orders of magnitude higher than those of monolithic aragonite. The inorganic and organic components in nacre have a high degree of organization not encountered in synthetic materials. From our analysis of toughening and strengthening mechanisms, we conclude that the unique structure of nacre is responsible for these superior properties. The current understanding of these toughening and strengthening mechanisms, and their relation to the sizes of the component phases, is far from complete.

Morphological and crystallographic analyses by electron microdiffraction of the inorganic phase biogenic aragonite indicate that the individual aragonite platelets in nacre form a multiple tiling system in which twins form hierarchical defect structures to control the overall structural order. Assuming that crystal-matrix recognition is applicable, a model is forwarded for the structural conformation of the active sites in the organic matrix which may then act as the template for the formation of nacre. This model, called the *superstructure*, explains all the experimentally observed crystallographic and morphological relationships in the aragonite phase. The geometric and crystallographic model of aragonite platelets proposed in this review is referred to as *multiple tiling*. It appears that nature utilizes this technique in nacre to form a highly ordered structure compatible with both the soft component and the crystalline and geometrical structural constraints of the hard component. Tiling may also play an important role in determining the overall shape formation of the nacre and its mechanical properties. Many mollusks species, such as gastropods, cephalopods,+ and bivalves have aragonite platelets as the fundamental building blocks in their nacre, but they also have grossly different overall shapes. In red abalone, for example, the shell is quite flat and thick; in nautilus, the shell is round and thin and forms an elegant chambered structure in which even the separation walls of the chambers are made of nacre. In all these nacre structures, the aragonite platelets have more or less the same dimensions, about 0.2 to 0.8 μm thick and 5 to 10 μm long on the edge. On the other hand, the multiple tiling of the platelets and the crystallographic relationship between the platelets may differ in these organisms due to slightly different structures and compositions in their underlying organic matrices. Further studies on various species of these organisms of both the crystallography of the mineral component and the structural and compositional analyses of the organic matrices, are warranted if we are to discern their structures and the unifying, underlying principles for the organization and formation of the various shapes of shell containing nacre structures. A fundamental understanding of these structures is essential for the possible formation of synthetic composites through biomimetics.

This new approach, in which the structure of an organic material is indirectly deduced from a detailed knowledge of the structural relationships among the subcomponents of the inorganic phase, may prove a viable approach for studying other biocomposites that also have a highly organized inorganic structure. This approach would serve as a new methodology, not only for studying the way the overall shape is determined in various species of mollusks, but also for understanding general biomineralization concepts in single and multicell organisms, including bone and teeth in higher organisms.

Despite considerable effort in the field, our understanding of the mechanisms that operate in nacre to create such a tough composite, and of the composition and structure of the organic matrix and its structural relationship with the inorganic phase are still limited. Some of the major issues, from which an agenda for future research may be drawn, are as follows:

(i) *On Mechanical Properties.* A quantitative understanding of the toughening and strengthening mechanisms in nacre is necessary because these mechanisms depend upon the structural relationships of the organic and inorganic phases. Mechanical property

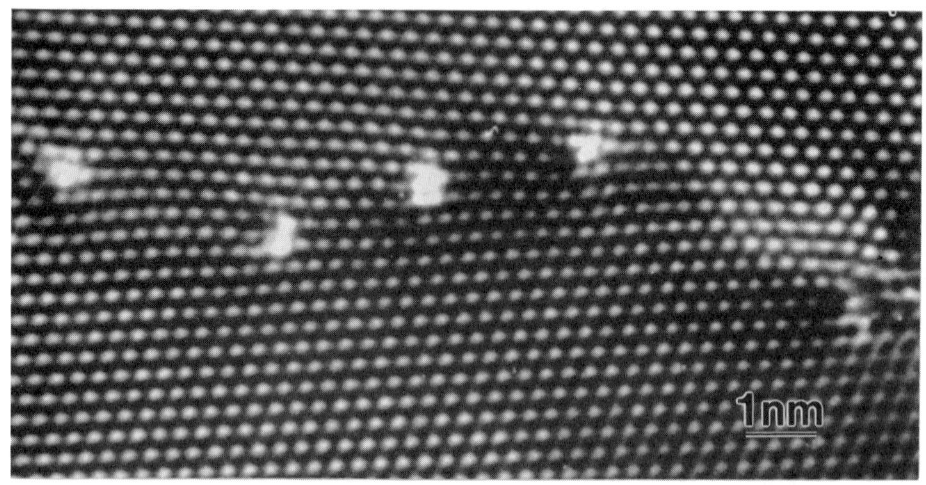

Figure 16. Atomic resolution transmission electron image of a low-angle boundary in biogenic aragonite in nacre of abalone displays dislocation core structures in the [001] electron beam direction (recorded at 400 kV). Although it is unusual for aragonitic lattice, as an ionic crystal, to have such defects, they are common substructural features in biogenic aragonite crystals.

evaluation of the overall shell, particularly under the dynamic conditions in which the organism lives and makes use of its multifunctional characteristics, is also essential for the design of multifunctional materials through biomimicking. Future research should include proper testing techniques for measurement including interfacial properties, analysis of paths for crack propagation, and their micromechanical analyses. For this, the knowledge of actual properties of the individual phases (organic matrix and biogenic aragonite) is essential, which, in turn, requires careful analysis of true structures [substructures and defects (Figure 16)] and compositions of the component phases. The coupling between the component phases and their size (size-effect) and conformation (crystallography) need to be studied to reveal sources of nanocomposite effects.

(ii) *On the Organic Matrix.* The protein and polysaccharide compositions in nacre, their conformations, identification as nucleator and framework macromolecules, and structural relationships of the organic matrix as well as its relationship to the inorganic aragonite phase are necessary parameters for consideration in the design of synthetic materials. The relationship between the organic matrix and calcite in the prismatic layer should also be investigated.

(iii) *On the Mechanism of Growth and Morphogenesis.* An investigation of the nucleation of the inorganic phase in the presence of an organic matrix, beginning at the embryonic and juvenile stages of the mollusk and comprehension of the degree to which the organic matrix controls growth will assist the understanding the overall formation and shaping of the shell.

(iv) *On Biomimicking.* New synthetic strategies are necessary to process thinner ceramic, metal, and polymeric layers for ceramic-metal and ceramic-organic composites with controlled substructures and interfaces, controlled lamination to achieve novel layered composites with desired architecture.

(v) *On Bioduplication.* Fundamental studies involving *in vitro* isolation, purification, and self-assembly of various components of the shell organic matrices, separately and in combination, are essential as the ground work. These would be followed by biomineralization of $CaCO_3$ and other biogenic ceramics on biogenic or engineered organic matrices under controlled conditions leading to a fundamental understanding of how the synthesis and structural evolution of hard biological structures might be controlled. Finally, these studies would open up new avenues in biduplication of technologically more significant ceramics, such as Fe_3O_4 and $BaTiO_3$, in various morphological forms (particles, platelets, laminates, and thin films).

Both biomimicking and, eventually, bioduplication may prove invaluable tools for materials engineers. Using these methods to design and process novel materials similar to biogenic multifunctional, nanolaminated composites like nacre, however, will be deferred until we answer the crucial questions discussed regarding the structure and function of this and other similar biocomposite systems.

7.0 ACKNOWLEDGEMENTS

This work was performed under the sponsorship of Air Force Office of Scientific Research under Grant # AFOSR-91-0281, and Army Research Office, Grant # DAAL03-92-G-0241.

8.0 REFERENCES

1. See, for instance, i. P. Haasen, *Physical Metallurgy* (Cambridge University Press, London, 1978); *Ceramic Microstructures*, R. M. Fulrath and J. A. Pask (eds.) (John Wiley and Sons, New York, 1968); ii. *High Performance Polymers*, E. Baer and A. Moet (eds.) (Hanser, Munich, 1991).

2. i. Nippon-to no kagateki kenkyu, K. Tawara (Tokyo, 1953); ii. H. Tanimura, "Development of the Japanese Sword," *JOM*, 32 [2] 63-73 (1980); iii. *A Search for Structure,* C. S. Smith (MIT Press, Cambridge, 1981) pp. 66-72; iv. O. D. Sherby, T. Oyama, D. W. Kum, and J. Wadsworth, *JOM*, **37** [6] 50-56 (1985).

3. R. P. Anders et al., "Research Opportunities on Clusters and Cluster Assembled Materials," *J. Mater. Res.*, **4**, 704-496 (1989).

4. See, for instance, *Decomposition of Austenite by Diffusional Processes*, V. F. Zackay and H. I. Aaronson (ed.) (Interscience, New York, 1962).

5. *Phase Transformations in Metals and Alloys*, D. A. Porter and K. E. Easterling (Van Nostrand Reinhold, Berkshire, UK, 1981) Chap. 5.

6. i. G. S. Kreimer, *Strength of Hard Alloys* (Consultants Bureau, New York, 1968) pp. 1-30; ii. H. E. Exner and J. Gurland, "A Review of Parameters Influencing Some Mechanical Properties of Tungsten Carbide - Cobalt Alloys," *Powder Metallurgy*, **13**, 13-31 (1970); iii. H. C. Lee and J. Gurland, "Hardness and Deformation of Cemented Tungsten Carbide," *Mater. Sci. Eng.*, **33**, 125-133 (1978).

7. *Superalloys*, M. Gell, D. H. Duhl, and A. F. Giamei (ASM-International, Metals Park, OH, 1980).

8. i. D. Baral, J. B. Ketterson, and J. E. Hilliard, "Mechanical Properties of Compositionally Modulated Cu-Ni Foils," *J. Appl. Phys.*, **57** [4] 1076-1084 (1985); ii. Anomalous Increase in the Strength of In-Situ Formed Ultrafine Cu-Nb Multifilamentary Composites," *J. Appl. Phys.*, **49** [12] 6031-6038 (1979); iii. *Metallic Superlattices,* T. Shinjo and T. Takada (eds.) (Elsevier, Amsterdam, 1987).

9. See, for instance, *High Temperature Superconductors: Relationships Between Properties, Structure, and Solid-State Chemistry*, J. D. Jorgensen, K. Kitazawa, J. M. Tarascon, M. S. Thomson, and J. B. Torrance (eds.) Proc. MRS, Vol. 156 (Materials Research Society, Pittsburgh, 1989).

10. J. F. Scott and C. A. Paz de Araujo, "Ferroelectric Memories," *Science*, **246**,1400-1405 (1989).

11. i. R. Bringer, U. Herr, and H. Gleiter, "Nanocrystalline Materials - A First Report," *Trans. Jap. Inst. Metall.*, Suppl., **27**, 43-52 (1986); ii. *Microclusters*, S. Sugano, Y. Nishina, and S. Ohnishi (eds.) (Springer and Verlag, Amsterdam, 1987); iii. L. E. Brus, "Semiconductor Crystallites," *Acc. Chem. Res.*, **23**, 183-188 (1990).

12. See, for instance, *Layered Structures, Epitaxy, and Interfaces*, J. M. Gibson and L. R. Dawson (eds.) Proc. MRS ,Vol. 37 (Materials Research Society, Pittsburgh, 1984).

13. See, for instance, *Epitaxial Heterostructures*," D. W. Shaw, J. C. Bean, V. G. Keramidas, and P. S. Peercy (eds.) Proc. MRS, Vol. 198 (Materials Research Society, Pittsburgh, 1989).

14. See, for instance, i. R. T. Bate, "The Quantum Size Effect: Tomorrow's Transistors," *Scientific American*, **258** [3] 96-106 (1988); ii. *Clusters and Cluster-assembled Materials*, R. S. Averback, D. L. Nelson, and J. Bernhole (eds.) Proc. MRS, Vol. 206 (Materials Research Society, Pittsburgh, 1991); iii. *Microclusters*, S. Ugano, Y. Nishina, and S. Ohuishi (eds.) (Springer and Verlag, Heidelberg, 1987).

15. *Smart Structures and Materials*, 1993 North American Conference, Albuquerque, New Mexico, U.S.A., Jan. 31 - Feb. 4, 1993.

16. R. E. Newnham and G. R. Rushan, "Smart Electroceramics," *J. Am. Ceram. Soc.*, **74** [3] 463-480 (1991).

17. See for instance, i. *Materials Research Bulletin*, **12** [3] (1987); ii. R. E. Newnham, S. E. McKinstry, and H. Ikawa, "Multifunctional Ferroic Nanocomposites," in: *Multifunctional Materials*, A. J. Buckley, G. Gallagher-Daggitt, F. E. Karasz, and D. R. Ulrich (eds.) Proc. MRS, Vol. 175 (Materials Research Society, Pittsburgh, 1990) pp. 161-168.

18. E. Baer, A. Hiltner, and R. J. Morgan, "Biological and Synthetic Hierarchical Composites," *Physics Today*, October, 60-67 (1992).

19. *Mechanical Design in Organisms*, S. A. Wainwright, W. D. Briggs, J. D. Currey, and J. M. Gosline (eds.) (John Wiley & Sons, New York, 1976).

20. J. D. Currey, "Biological Composites," *J. Mater. Edu.* **9** [1-2] 118-296 (1987).

21. i. *Biomineralization*, K. Simkiss, and K. M. Wilbur (Academic Press, New York, 1989); ii. *On Biomineralization*, H. A. Lowenstam and S. Weiner (Oxford University Press, New York, 1989); iii. *Biomineralization: Chemical and Biochemical Perspectives*, S. Mann, J. Webb, and R. J. Williams (eds.) (VCH Publishers, Weinheim, 1989).

22. i. P. Calvert and S. Mann, "Synthetic and Biological Composites Formed by *in situ* Processing," *J. Mater. Sci.*, **23**, 3801-3815 (1988); ii. *Materials Synthesis Using Biological Processes*, P. C. Rieke, P. D. Calvert, and M. Alper (eds.), Proc. MRS, Vol. 174 (Materials Research Society, Pittsburgh, 1990); iii. *Materials Synthesis Based on Biological Process*, M. Alper, P. D. Calvert, R. Frankel, P. Rieke, and D. Tirrell (eds.) Proc. MRS, Vol. 218 (Materials Research Society, Pittsburgh, 1991); iv. *Hierarchically Structured Materials*, I. A. Aksay, E. Baer, M. Sarikaya, and D. A. Tirrell (eds.), Proc. MRS, Vol. 255 (Materials Research Society, Pittsburgh, 1992).

23. *Biochemistry*, C. K. Mathews and K. E. van Holde (The Benjamin/Cummings Publ. Co., Redwood City, CA, 1990).

24. H. A. Lowenstam, "Biological Minerals," *Science*, **211**, 1126-1128 (1981).

25. T. Degens, "Molecular Mechanisms on Carbonate, Phosphate, and Silica Deposition in the Living Cell," *Top. Curr. Chem.*, **64**, 1-112 (1976).

26. i. M. Paulsson, "Basement Membrane Proteins: Structure, Assembly, and Cellular Interactions," *Critic. Rev. in Biochem. and Molec. Bio.*, **27** [1/2] 93-120 (1992); ii. W. A. Cramer, D. M. Engleman, and G. von Heijne, "Forces Involved in the Assembly and Stabilization of Membrane Proteins," *FASEB J.*, **6** [15] 3397-3405 (1992); iii. T. Hashimoto, K. Kimishama, and H. Hasegawa, "Self-Assembly and Patterns in Binary Mixtures of SI Block Copolymers and PPO," *Macromol.*, **24** [20] 5604-5710 (1991).

27. L. Addadi and S. Weiner, "Interaction between Acidic Proteins and Crystals: Stereochemical Requirements in Biomineralization," *Proc. Natl. Acad. Sci., USA, Biophysics*, **82**, 4110-4114 (1985).

28. i. H. R. Crane, "Principles and Problems of Biological Growth," *Sci. Monthly*, **70**, 376-389 (1950); ii. S. Mann, "Mineralization in Biological Systems," *Structure and Bonding*, Vol. 54 (Springer-Verlag, Berlin, 1983) pp. 125-174.

29. K. M. Wilbur, "Shell Formation and Regeneration," in *Physiology of Mollusca*, K. M. Wilbur and C. M. Yonghe (eds.) Vol. 1 (Academic Press, New York, 1964) p. 243.

30. M. Sarikaya and I. A. Aksay, "Nacre of Abalone Shell: a Natural Multifunctional Nanolaminated Ceramic-Polymer Composite Material," in: *Results and Problems in Cell Differentiation in Biopolymers*, Steven Case (ed.) (Springer and Verlag, Amsterdam, 1992) pp. 1-25.

31. L. Addadi and S. Weiner, "Interactions between Acidic Macromolecules and Structured Surfaces; Stereochemistry and Biomineralization," *Mol. Cryt. Liq. Cryst.*, **134**, 305-322 (1986).

32. Y. Bouligand, "Sur une architecture torsadee repandue dans de nombreuses cuticules d'arthropodes," *C R Hebd. Sceances Acad. Sci.*, **261** [12] 3665-3668 (1965).

33. M. M. Giraud-Guille, "Fine Structure of the Chitin-Protein System in the Crab Cuticle," *Tissue Cell* **16**, 75-92 (1984).

34. *Iron Biominerals*, B. Frankel and R. P. Blakemore (eds.) (Plenum, New York, 1991).

35. M. Gosline, M. E. DuMont, and M. W. Denny, "Structure and Properties of Spider Silk," *Endeavour* **10** [1] 37-43 (1986).

36. M. Denny and J. M. Gosline, "The Physical Properties of the Terrestrial Slug," *J. Exp. Biol.*, **88**, 375-393 (1980).

37. A. I. Caplan, "Cartilage," *Sci. Amer.*, **251** [4] 84-94 (1984).

38. i. *Biochemistry of Collagen*, G. N. Ramachandran and R. H. Reddi (eds.) (Plenum Press, New York, 1976); ii. Y. Bouligand and M. M. Giraud-Guille, "Spatial Organization of Collagen Fibrils in Skeletal Tissues: Analogies with Liquid Crystals," in: *Biology of Invertebrate Collagens*, A. Bairati and R. Garrone (eds.) (Plenum, New York, 1985) pp. 115-134; iii. E. Baer, J. J. Cassidy, and A. Hiltner, "Hierarchical Structure of Collagen and Its Relationship to the Physical Properties of Tendon," Chap. 9, in: *Collagen: Biochemistry and Biomechanics*, M. E. Nimi (ed.) (CRC Press, Inc., New York, 1988) pp. 177-199; iv. P. Borstein and W. Traub, "The Chemistry and Biology of Collagen," in *The Proteins*, H. Neurath and R. L. Hill (eds.) 3rd Ed. (Academic Press, New York, 1979) pp. 412-632.

39. W. M. Kier and K. K. Smith, "Tangues, Tentacles, and Trunks: Biomechanics of Movement in Muscular Hydrostats," *Zool. J. Linn. Soc.*, **83**, 307-324 (1985).

40. *Introduction to Plant Biochemistry*, T. W. Goodwin and E. I. Mercer (Pergamon, Oxford, 1983).

41. See, for instance, References 20, 21, 25, and 30.

42. See, for instance, A. Boyde, "Comparative Histology of Mammalian Teeth," in *Dental Morphology and Evolution*, A. A. Dahlberg (ed.) (Chicago University Press, Chicago, 1971) pp. 81-94. Also, see Reference 43.

43. *Chemistry and Biology of Mineralized Tissues*, H. C. Slavkin and P. Price (eds.) (Excerpta Medica, Elsevier, Amsterdam, 1992).

44. K. Markel and P. Gorny, Zur Funktionellen Anatomie der Seeigelzahne (Echinodermata and Echinoidea), *Z. Morphol. Tiere*, **75**, 223-242 (1973).

45. K. M. Towe, "Echinoderm Calcite: Single Crystal or Polycrystalline Aggregate, "*Science*, **157**, 1048-1050 (1967).

46. i. R. P. Blakemore, "Magnetotactic Bacteria," *Science* **190**, 377-379 (1975); ii. R. P. Blakemore, "Magnetotactic Bacteria," *Ann. Rev. of Microbiology*, **36**, 217-238 (1982).

47. C. T. Dameron, R. N. Reese, R. K. Mehra, A. R. Kortan, P. J. Carrol, M. L. Steigerwald, L. E. Brus, and D. R. Winge, "Biosynthesis of Cadmium Sulfide Quantum Semiconductor Crystallites," *Nature*, **338**, 596-597 (1989).

48. i. *Enzyme Structure and Mechanism*, A Fersht, 2nd Ed. Chaps. I and II (Freeman, New York, 1985); ii. M. Alper, "Enzymatic Synthesis of Materials - An Overview," in *Materials Synthesis Based on Biological Processes*, M. Alper, P. Calvert, R. B. Frankel, P. Rieke, and D. A. Tirrell (eds.) Proc. MRS, Vol. 218 (Materials Research Society, Pittsburgh, 1991) pp. 3-6.

49. S. F. Mathews, "The Structure, Function, and Evolution of Proteins," *Prog. Biophys. Mol. Sci.*, **45**, 1-56 (1985).

50. M. Fournier, H. S. Creel, K. P. McGrath, M. T. Krejchi, E. D. T. Atkins, T. L. Mason, and D. A. Tirrell, "Genetic Synthesis of Periodic Protein Materials," J. *Bioct. Compat. Polymers*, **6**, 326-338 (1991).

51. *Research Opportunities for Materials with Ultrafine Microstructures*, National Research Council (National Academy Press, Washington, DC, 1989).

52. P. M. Harrison, P. J. Artymiuk, G. C. Ford, D. M. Lawson, J. M. A. Smith, A. Treffry, and J. L. White, "Ferritin: Function and Structural Design of an Iron-Storage Protein," in *Biomineralization: Chemical and Biochemical Perspectives*,

edited by S. Mann, J. Webb, and R. J. P. Williams (eds.) (VCH, Weinheim, 1989) pp. 257-294.

53. i. B. Frankel, R. P. Blakemore, and R. S. Wolfe, "Magnetite in Freshwater Magnetotactic Bacteria," *Science*, **203**, 1355-1356 (1979); ii. R. P. Blakemore, "Magnetotactic Bacteria," *CRC Critical Rev. in Biochem.*, **20** [4] 365-380 (1986).

54. D. A. Bazylinski, R. B. Frankel, A. Garrat-Reed, and S. Mann, "Biomineralization of Iron Sulfides in Magnetotactic Bacteria from Sulfidic Environments," in *Iron Biominerals*, R. B. Frankel and R. P. Blakemore (eds.) (Plenum Press, New York, 1991) pp. 239-256.

55. i. B. M. Moskowitz, R. B. Frankel, P. J. Flanders, R. P. Blakemore, and B. B. Schwartz, "Magnetic Properties of Magnetotactic Bacteria," *J. Magnetism and Mag. Maters.*, **73**, 273-280 (1988); ii. S. Krieger, G. J. Olson, J. J. Rhyne, R. P. Blakemore, Y. A. Gorby, and N. Blakemore, "Small Angle Neutron and X-Ray Scattering from Magnetic Crystals in Magnetotactic Bacteria," *J. Magnetism and Mag. Maters.*, **82**, 17-28 (1989); iii. K. M. Towe and T. T. Moench, "Electron-Optical Characterization of Bacterial Magnetite," *Earth and Planetary Sci. Lett.*, **52**, 213-220 (1981).

56. i. S. Mann, R. B. Frankel, and R. P. Blakemore, "Structure, Morphology, and Crystal Growth of Bacterial Magnetite," *Nature*, **310**, 405-407 (1984); ii. S. Mann, N. H. C. Sparks, and V. J. Wade, "Crystallochemical Control of Iron Oxide Biomineralization," in *Iron Biominerals*, R. B. Frankel and R. P. Blakemore (eds.) (Plenum Press, New York, 1990).

57. M. Sarikaya, N. Pellerin, J. T. Staley, and I. A. Aksay, unpublished results (1992).

58. J. F. Stolz, S-B R. Chang, and J. L. Kirschvink, "Magnetotactic Bacteria and Single-Domain Magnetite in Hemipelagic Sediments," *Nature*, **321**, 849-851 (1986).

59. W-H. Shih, M. Sarikaya, W. Y. Shih, and I. A. Aksay, "Geometrical Arrangement of Magnetosomes in Magnetotactic Bacteria," in *Materials Synthesis Based on Biological Processes*, M. Alper, P. Calvert, R. B. Frankel, P. Rieke, and D. A. Tirrell (eds.) Proc. of MRS, Vol. 218 (Materials Research Society, Pittsburgh, 1991) pp. 109-114.

60. i. See paper by D. A. Bazylinski and R. B. Frankel, "Biomineralization of Iron Sulfides in Magnetotactic Bacteria in Sulfidic Environments," in ref. 34., pp. 239-255, and also D. A. Bazylinski, A. J. Garratt-Reed, and R. B. Frankel, "Electron Microscopic Studies in Magntetotactic Bacteria, in: *Microscopy of Self-Assembled*

Materials and Biomimetics, M. Sarikaya (ed.) *J. Microsc. Res. Tech* (1993); ii. M. Farina, D. M. S. Esquivel, and H. G. P. Lins de Barros, "Magnetic Iron-Sulphur Crystals from Magnetotactic Microorganisms," *Nature*, **343**, 256-258.

61. i. *Membranemimetic Chemistry*, J. H. Fendler (Wiley-Interscience, New York, 1982); ii. J. H. Fendler, "Atomic and Molecular Clusters in Membrane Mimetic Chemistry," *Chem. Rev.*, **87**, 887-899 (1987).

62. i. H. Liu, G. L. Graff, M. Hyde, M. Sarikaya, and I. A. Aksay, "Synthesis of Ultrafine Multicomponent Particles Using Phospolipid Vesicles," in *Materials Synthesis Based on Biological Processes*, M. Alper, P. Calvert, R. B. Frankel, P. Rieke, and D. A. Tirrell (eds.) Proc. MRS, Vol. 218 (Materials Research Society, Pittsburgh, PA, 1991) pp. 115-121; ii. Sung Pak, "Potential Application of Microemulsions in Ceramic Processing," M.S. Thesis (University of Washington, Seattle, 1979).

63. I. A. Aksay, W. Y. Shih, and M. Sarikaya, "Colloidal Processing of Ceramics with Ultrafine Particles," in *Ultrastructure Processing of Advanced Ceramics, Glasses, and Composites*, J. D. MacKenzie and D. R. Ulrich (eds.) (Wiley, New York, 1988) pp. 393-406.

64. Y. A. Gorby, T. J. Beveridge, and R. P. Blakemore, "Characterization of the Bacterial Magnetosome Membrane," *J. Bacteriology*, **170** [2] 834-841 (1988).

65. R. B. Frankel and D. A. Bazylinski, "Structure and Function of Magnetosomes in Magnetotactic Bacteria," this book, pp. 199-216.

66. E. Kniprath, "Ultrastructure and Formation of Sea Urchin Tooth," *Calcif. Tiss. Res.*, **14**, 211-228 (1974).

67. i. S. Weiner, "Organic Matrixlike Macromolecules Associated with the Mineral Phase of Sea Urchin Skeletal Plates and Teeth" *J. Exp. Zool.*, **234**, 7-15 (1985); ii. D. J. Veis, T. M. Albinger, J. Clohisy, M. Rahima, B. Sabsay and A. Veis, "Matrix Proteins of Sea Urchin *Lytechinus variegatus*," *J. Exp. Zool.*, **240**, 35-46 (1986); iii. A. Berman, L. Addadi, L. Leiserowitz, S. Weiner, M. Nelson, and A. Kvik, "A Synchrotron X-ray Study of a Unique Protein-Calcite Composite Material," *Science*, **250**, 664-667 (1990).

68. M. Sarikaya, J. Liu, and I. A. Aksay, "Ultrastructure of the Mineral Phase in Sea-Urchin Teeth," unpublished research.

69. K. Brear and J. D. Curry, "Structure of the Sea-Urchin Tooth," *J. Mater. Sci.*, **11**, 1977-1978 (1976).

70. I. A. Aksay and M. Sarikaya, "Bioinspired Processing of Composite Materials" in *Ceramics: Toward the 21st Century, Centennial International Symposium,* S. Soga and A. Kato (eds.), (Ceramic Society Japanese, Tokyo, 1991) pp. 136-149.

71. *Fundamental Principles of Fiber Reinforced Composites*, K. H. G. Ashbee (Technomic Publishing, Lanchester, PA, 1989).

72. J. Kastelic, I. Palley, and E. Baer, "A Structural Mechanical Model for Tendon Crimping," *J. Biomechanics*, **13**, 887-893 (1980); ii. E. Baer, J. J. Cassidy, and A. Hiltner, "Hierarchical structure of Collagen Composite Systems: Lessons from Biology," this book, pp. 13-34.

73. i. D. L. Kaplan, S. J. Lombardi, W. S. Muller, and S. A. Fossey, "Silks," in *Biomaterials*, D. Byrom (ed.) (Stockton Press, New York, 1991) pp. 1-53; ii. J. Gosline, C. Nichols, P. Guerette, A. Cheng, and S. Katz, "The Macromolecular Design of Spiders' Silks," this book, pp. 237-262.

74. *The Biology of Anthropod Cuticle*, A. Neville (Springer, Berlin, 1975).

75. M. M. Giraud-Guille, "Liquid Crystalline Order of Biopolymers in Cuticles and Bones," in *Microscopy of Self-Assembled Materials and Biomimetics*, M. Sarikaya (ed.) Special Issue of *J. Microsc. Res. Tech.* (1994) pp. 420-428.

76. i. S. L. Gunderson and R. C. Schiavone, "The Insect Exoskeleton: A Natural Structural Material," *JOM*, **41** [11] 80-82 (1989); ii. S. L. Gunderson and R. C. Schiavone, "Microstructure of an Insect Cuticle and Applications to Advanced Composites," this book, pp. 163-198.

77. H. Ghiradella, D. Aneshansley, T. Eisner, R. E. Silberglied, and H. E. Hinton, "Ultraviolet Reflection of Male Butterfly: Interference Color Caused by Thin-Layer Elaboration of Wing Scales, *Science*, **178**, 1214-1217 (1972).

78. i. J. Samseth, R. J. Spontak, and K. Mortensen, "The Response of Microstructure to Processing in a Series of Poly (siloxaneimide) Copolymers," *J. Poly. Sci.*, Part B, **31** [4] 467-475 (1993); ii. R. J. Spontak, S. D. Smith, and A. Ashraf, "Morphological Studies of Linear $(AB)_n$ Multiblock Copolymers and Their Blends," in *Microscopy of Self-Assembled Materials and Biomimetics*, M. Sarikaya (ed.) Special Issue of *J. Microsc. Res. Tech.* (1994) pp. 412-419.

79. E. Baer, "Advanced Polymers," *Scientific American*, **255** [4] 178-190 (1986).

80. J. Glimcher, "On the Form and Function of Bone: from Molecules to Organs" in *The Chemistry and Biology of Mineralized Biological Tissues: Wolff's Law Revisited,* A. Veis (ed.) (Elsevier, New York and Amsterdam, 1981) pp. 617-673.

81. *Studies in the Development, Function, and Evolution of Teeth*, P. M. Butler and K. A. Joysey (eds.) (Academic, London, 1973).

82. C. Grégoire, "Structure of the Molluscan Shell," in: *Chem. Zoology,* M. Florkin and M. Scheer (eds.) (Academic Press, New York, 1972) pp. 45-102.

83. i. J. D. Currey, "The Mechanical Properties of Some Molluscan Hard Tissues," *J. Zool. London,* **173**, 39-406 (1974); ii. J. D. Currey, "Further Studies on the Mechanical Properties of Mollusc Shell Material," *J. Zool. London,* **180**, 445-453 (1976).

84. A. P. Jackson, J. F. V. Vincent, and R. M. Tunner, "The Mechanical Design of Nacre," *Proc. Roy. Soc. London*, **B234**, 415-440 (1988)

85. M. Sarikaya, K. E. Gunnison, M. Yasrebi, and I. A. Aksay, "Mechanical Property-Microstructural Relationships in Abalone Shell," *Materials Synthesis Using Biological Processes*, P. C. Rieke, P. D. Calvert, and M. Alper (eds.) Proc. MRS, Vol. 174 (Materials Research Society, Pittsburgh, 1990) pp. 109-116.

86. W. Schober, "Precipitated Calcium Carbonate: A Quite Market Expects Excellence," *Industrial Minerals*, No. 265, October Issue, 69-81 (1989).

87. i. H. Nakahara, G. Bevelander, and M. Kakei, "Electron Microscopic and Amino Acid Studies on the Outer and Inner Shell Layers of *Haliotis rufescens*," *VENUS* (*Japn. Jour. Malac.*) **41** [1] 33-46 (1982); ii. G. Bevelander and H. Nakahara, "An Electron Microscopy Study of the Formation of the Nacreous Layers in the Shells of Certain Bivalve Molluscs," *Calc. Tiss. Res.,* **3**, 84-92 (1968).

88. i. K. Bandel, "Ubergange von der Perlmutter-Schicht zu Prismatischen Schichttypen bei Mollusken," Biomineralization, **2**, 28-47 (1970); ii. K. M. Wilbur and K. Simkiss, "Calcified Shells," *Comprehensive Biochemistry*, **26A,** 229-295 (1968).

89. N. Watabe, "Studies on Shell Formation: XI. Crystal-Matrix Relationship in the Inner Layers of Mollusk Shells," *Ultrastr. Res.*, **12**, 351-370 (1965).

90. i. S. Weiner, W. Traub, and H. A. Lowenstam, "Organic Matrix in Calcified Exoskeletons," in *Biomineralization and Biological Metal Accumulation*, P. Westbroek and E. W. de Jong (eds.) (Dordrecht: Reidel, 1983) pp. 205-224; ii. S.

Weiner, Y. Talmon, and W. Traub, "Electron Diffraction Studies of Molluscan Shell Organic Matrices and Their Relationship to the Mineral Phase," *Int. J. Biol. Macromol.* **5,** 325-328 (1983); iii. S. Weiner and W. Traub, "Macromolecules in Mollusc Shells and Their Functions in Biomineralization," *Phil. Trans. R. Soc. Lond.*, B **304**, 425-434 (1984); iv. D. Worms and S. Weiner, "Mollusk Shell Organic Matrix: Fourier Transform Infrared Study of the Acidic Macromolecules," *J. Exp. Zool.*, **237**, 11-20 (1986);

91. *The Biology of Marine Animals,* A. Nicol (Interscience, New York, 1960).

92. *Molluscs,* E. Morton, 5th Edition (Hutchinson, London, 1979).

93. i. H. Mutvei, "Ultrastructure of the Mineral and Organic Components of Molluscan Nacreous Layers," *Biomineralization* **2**, 48-72 (1970); ii. H. Mutvei, "Ultrastructural Characteristics of the Nacre in Some Gastropods," *Zool. Scripta*, **7**, 287-296 (1978).

94. *Physiology of Mollusca,* K. Wilbur and C. M. Yonge (eds.) (Academic Press, New York, 1964).

95. i. W. F. Brown, Jr. and J. E. Srawley, "Plain Strain Fracture Toughness Testing of High Strength Metallic Materials," *ASTM Technical Publ. No 410* (American Society for Testing and Materials, Philadelphia, 1966); ii. *Plane Strain Fracture Toughness Testing*, ASTM E599 (American Society for Testing and Materials, Philadelphia, 1983).

96. J. D. Currey, "Mechanical Properties of Mother of Pearl in Tension," *Proc. Roy. Soc. London*, B**196**, 443-463 (1977).

97. *Deformation of Ceramic Materials II,* R. E. Tressler and R. C. Bradt (eds.) *Materials Science Research*, Vol. 18 (Plenum Press, New York, 1984).

98. *Science and Technology of Zirconia,* N. Claussen, M. Ruhle, and A. Heuer (eds.) (American Ceramic Society, Inc., Columbus, Ohio, 1984); ii. S. M. Wiederhorn, "Brittle Fracture and Toughening Mechanisms in Ceramics," *Ann. Rev. Mater. Sci.*, **14**, 373-403 (1984).

99. See, for instance, W. Davidge, *Mechanical Behaviour of Ceramics* (Cambridge University Press, Cambridge, 1979).

100. See, for instance, i. References 30 and 85, and also, ii. M. Yasrebi, G. H. Kim, D. L. Milius, M. Sarikaya, and I. A. Aksay, "Biomimetic Processing of Ceramics and Ceramic-Based Composites," in: *Better Ceramics Through Chemistry-IV*, B. J. J.

Zelinski, C. J. Brinker, D. E. Clark, and D. R. Ulrich (eds.) Proc. MRS, Vol. 180 (Materials Research Society, Pittsburgh, 1990) pp. 625-635.

101. See, for instance, *The Plastic Deformation of Metals*, W. K. Honeycombe (St. Martin's Press, New York, 1968).

102. A. A. Griffith "The Phenomena of Rupture and Flow in Solids," *Phil. Trans. CCXXI* **A**, 163-198 (1920).

103. *An Introduction to Composite Materials*, D. Hull (Cambridge University Press Cambridge, 1981) pp. 81-85.

104. i *British and Continental Arms and Armor*, C. H. Ashdown (Dover Publications, New York, 1975); ii. *Arrows Against Steel: The History of the Bow*, (Mason/Charter, New York, 1975); iii. V. Hurley, M. L. Wilkins, C. F. Cline, and C. A. Honodel, *Light Armor*, Report UCRL-7-1817, Lawrence Livermore National Laboratory, Livermore, CA, USA (1969).

105. H. Silyn-Robertson and R. M. Sharp, "Crystal Growth and the Role of the Organic Network in Eggshell Biomineralization," *Proc. Roy. Soc. Lond.*, **B 227**, 303-324 (1986).

106. R. W. Rice, "Microstructure Dependence of Mechanical Behavior," in *Treatise on Materials Science and Technology*, Vol. II, *Properties and Microstructure*, R. K. McCrone (ed.) (Academic Press, New York, 1977) pp. 199-381.

107. i. A. G. Evans and R. M. MacMeeking, "On the Toughening of Ceramics by Strong Reinforcements," *Acta. Metall.*, **34** [12] 2435-2441 (1986); ii. A. G. Evans, "High Toughness Ceramics," *J. Mater. Sci. Eng.*, **A105/106**, 65-75 (1988); iii. A. G. Evans and D. A. Marshall, "The Mechanical Behavior of Ceramic Matrix Composites," *Acta. Metall.* **37** [10] 2567-2583 (1989).

108. S. Khanuja, "Processing and Structure-Property Relationships of Laminated B_4C-Polymer Composites," *M.Sc. Thesis* (University of Washington, Seattle, 1991).

109. G. H. Kim, "The Effect of Metallic Phase and Microstructure on the Strengthening Behavior of B_4C-Al Cermets, *Ph.D. Thesis* (University of Washington, Seattle, 1993).

110. See, for instance, i. P. Boch, T. Chartier, and M. Huttepain, "Tape Casting of Al_2O_3/ZrO_2 Laminated Composites," *J. Am. Ceram. Soc.*, **69** [8] C191-C192 (1986); ii. H. Takeba and K. Morinaga, "Fabrication and Mechanical Properties of Lamellar Al_2O_3 Ceramics," *J. Ceram. Soc. Jap. Int'l. Ed.*, **96**, 1122-1128 (1988);

iii. D. B. Marshall, J. J. Ratto, and F. F. Lange, "Enhanced Fracture Toughness in Layered Nanocomposites of Ce-ZrO_2 and Al_2O_3," *J. Am. Ceram. Soc.*, **74** [12] 2979-2987 (1991).

111. i. K. M. Towe and G. H. Hamilton, "Ultrastructure and Inferred Calcification of the Mature and Developing Nacre of Bivalve Mollusks," *Calc. Tiss. Res.*, **1**, 306-318 (1968); ii. K. M. Towe and G. R. Hamilton, "The Structure of Some Bivalve Shell Carbonates Prepared by Ion-Beam Thinning," *Calc. Tiss. Res.*, **10**, 38-48 (1972).

112. i. Bragg, "The Structures of Aragonite," *Proc. Roy. Soc. London, Ser. A*, **17** (1928); ii R. Wenk, D. J. Barber, and R. J. Reeder, "Microstructures in Carbonates" in *Reviews in Mineralogy*, Vol. 11, R. J. Reeder (ed.) (Miner. Soc. Amer., Washington D.C., 1983) p. 301.

113. M. A. Crenshaw and H. Ristedt, "Histochemical and Structural Study of Nautiloid Septal Nacre," *Bomineralization*, **8**, 1-8 (1975).

114. i. S. Weiner, "Organization of Extracellularly Mineralized Tissues: A Comparative Study of Biological Crystal Growth," *CRC Crit. Rev. in Biochem.*, **20** [4] 365-380 (1986); ii. L. Addadi and S. Weiner, "Interactions between Acidic Macromolecules and Structured Crystal Surfaces: Stereochemistry and Biomineralization," *Mol. Cryst. Liq. Cryst.* **134,** 305-322 (1986); L. Addadi and S. Weiner, "Control and Design Principles in Biological Mineralization," *Angew. Chem. Intl. Ed. Engl.*, **31**, 153-169 (1992).

115. S. Mann, "Molecular Recognition in Biomineralization," *Nature*, **33**, 119-123 (1988).

116. J. A. Keith, S. A. Stockwell, D. H. Ball, W. S. Muller, D. L. Kaplan, T. W. Thannhauser, and R. W. Sherwood, "Characterization of the Complex Matrix of the Mytilus Edulis Shall and the Implications for Biomimetic Ceramics," in *Hierarchically Structured Materials*, I. A. Aksay, E. Baer, M. Sarikaya, and D. A. Tirrell (eds.), Proc. MRS, Vol. 255 (Materials Research Society, Pittsburgh, 1992) pp. 3-8.

117. M. A. Cariolou and D. E. Morse, "Purification and Characterization of Calcium-binding Conchiolin Shell Peptides from the Mollusc, *Haliotis rufescens*, as a Function of Development," *J. Comp. Physiol B.*, **157**, 717-729 (1988).

118. i. N. Watabe, "Crystal Growth of Calcium Carbonate in the Invertabrates," *Prog. Crystal Growth Charact.*, Vol. 4 (Pergamon Press, London, 1981) pp. 99-147; ii. C. Grégoire, "Structure of the Molluscan Shell," *Chem. Zool.*, **45**, 102-130 (1972);

S. W. Wise, "Microarchitecture and Mode of Formation of Nacre (mother-of-pearl) in Pelecypods, Gastrapods, and Cephalopods," *Eclogae geal Helv.*, **63** [3] 775-797 (1970).

119. i. M. Greenfield, D. C. Wilson, and M. A. Crenshaw, "Ionotropic Nucleation of Calcium Carbonate by Molluscan Matrix," *Amer. Zool.*, **24**, 925-932 (1984); ii. A. Crenshaw, "Mechanism of Normal Biological Mineralization of Calcium Carbonates" in *Biological Mineralization and Demineralization,* G. H. Nancollas (ed.) (Springer, Berlin and New York, 1972) pp. 243-257.

120. i. S. Weiner and W. Traub, "X-ray Diffraction Study of the Insoluble Organic Matrix of Mollusc Shells," *FEEBS Letters*, **111** [2] 311-316 (1980); ii. S. Weiner and W. Traub, "Organic Matrix-Mineral Relationships in Mollusc Shell Nacreous Layers" in *Structural Aspects of Recognition and Assembly in Biological Macromolecules*, M. Balaban, J. L. Sussman, W. Traub, and A. Yonath, (eds.) (Balaban ISS, Yehevot Philadelphia, 1981) pp. 462-487; iii. S. Weiner and W. Traub, "Macromolecules in Mollusc Shells and Their Functions in Biomineralization," *Phil. Trans. R. Soc. London* **B304,** 425-434 (1984).

121. i. D. E. Morse, University of California, Santa Barbara, CA - private communication (1992); ii. D. L. Kaplan, Army Research Center, Nattick, MA, private communication (1992); iii. C. E. Furlong, University of Washington, private communication (1993).

122. C. E. Furlong and R. Humbert, "Design of Protein Producing Bioreactors for Self-Assembling Systems," in *Hierarchically Structured Materials*, I. A. Aksay, E. Baer, M. Sarikaya, and D. A. Tirrell (eds.), Proc. MRS, Vol. 255 (Materials Research Society, Pittsburgh, 1992) pp. 435-442.

123. C. E. Furlong, R. Humbert, M. Sarikaya, and I. A. Aksay, unpublished research (1993).

124. K. Gunnison, M. Sarikaya, J. Liu, and I. A. Aksay, "Structure-Mechanical Property Relationships in a Biological Ceramic-Polymer Composite: Nacre," in *Hierarchically Structured Materials*, I. A. Aksay, E. Baer, M. Sarikaya, and D. A. Tirrell (eds.), Proc. MRS, Vol. 255 (MRS, Pittsburgh, 1992) pp. 171-183.

125. N. Watabe and K. M. Wilbur, "Influence of the Organic Matrix on Crystal type in Molluscs," *Nature*, **188,** 334-336 (1960).

126. R. Meenakshi, G. Donnay, P. L. Blackwelder, and K. M. Wilbur, "The Influence of Substrata on Calcification Patterns in Molluscan Shell," *Calc. Tiss. Res.*, **15**, 31-44 (1974).

127. A. Crenshaw, "Mechanism of Normal Biological Mineralization of Calcium Carbonates" in *Biological Mineralization and Demineralization,* G. H. Nancollas (ed.) (Springer, Berlin and New York, 1972) pp. 243-257.

128. S. Weiner, "Organization of Extracellularly Mineralized Tissues: a Comparative Study of Biological Crystal Growth," *CRC Crit. Rev. in Biochem*, **20** [4] 365-380 (1986).

129. S. Mann, "Biogenic Inorganic Materials," in *Inorganic Materials,* D. W. Bruce and D. O'Hare (eds.) (John Wiley & Sons, New York, 1992) pp. 237-294.

130. i. L. Addadi, Z. Berkovitch-Yellin, I. Weissbuch, M. Lahav, and L. Leiserowitz, "A Link between Macroscopic Phenomena and Molecular Chirality: Crystals as Probes for the Direct Assignment of Absolute Configuration of Chiral Molecules," in *Topics in Stereochemistry,* Vol. 16 (Wiley, New York, 1986) pp. 1-85; ii. I. Weisbuch, L. Addadi, M. Lahav, and L. Leiserowitz, "Molecular Recognition at Crystal Interfaces," *Science*, **253**, 637-645 (1991).

131. i. S. Mann, B. R. Heywood, S. Rajam, and D. Birchall, "Controlled Crystallization of $CaCO_3$ under Stearic Acid Monolayer," *Science*, **334,** 692-695 (1988); ii. S. Mann, B. R. Heywood, S. Rajam, and J. D. Birchall, "Interfacial Control of Calcium Carbonate Under Organized Stearic Acid Monolayers," *Proc. R. Soc. Lond.*, A, **423**, 457-471 (1989); iii. S. Mann, B. R. Heywood, S. Rajam, and J. B. A. Walker, "Structural and Stereochemical Relationships between Langmuir Monolayers and Calcium Carbonate Nucleation," *J. Phys. D: Appl. Phys.*, **24**, 154-164 (1991).

132. *Crystal Structures*, R. W. G. Wyckoff (Interscience Publ., New York, 1960).

133. See, for instance, D. Romeu, "Decagonal Phase Model with Multiple Periods and Quasiperiodic Coincidence Lattice," *J. Non-Cryst. Solids*, **153/154**, 232-240 (1993).

134. J. Liu, M. Sarikaya, and I. A. Aksay, "A Hierarchically Structured Model Composite: A TEM Study of the Hard Tissue of Red Abalone," in *Hierarchically Structured Materials*, I. A. Aksay, E. Baer, M. Sarikaya, and D. A. Tirrell (eds.), Proc. of MRS Symp., Vol. 258: (Materials Research Society, Pittsburgh, 1992) pp. 9-17.

135. N. Unwig and R. Henderson, "The Structure of Proteins in Biological Membranes," *Scientific American*, February, 78-94 (1985).

136. i. *Pattern and Tiling*, B. Grünbaum (Plenum, New York, 1987); ii. *Tilings and Patterns,* B. Grünbaum and G. C. Shephard (W. H. Freeman and Company, New York, 1987).

137. M. Sarikaya, J. Liu, and I. A. Aksay, unpublished research (1993).

138. *On the Growth and Form*, d'Arcy Thompson (University Press, Cambridge, 1952) Chap. XI.

139. D. R. Fowler, H. Meinhardt, and P. Prusinkiewicz, "Modelling Seashells," *Computer Graphics*, **26** [2] 379-387 (1992).

BIOMINERALIZATION, THE INORGANIC-ORGANIC INTERFACE, AND CRYSTAL ENGINEERING

Stephen Mann

School of Chemistry, University of Bath
Bath BA2 7AY, United Kingdom

The study of biomineralization and related model systems has advanced to a level where major insights in materials science can be expected in the near future. The application of principles such as molecular recognition and self-assembly in the regulation of nucleation and growth is pivotal to a new perspective in the fabrication of advanced inorganic materials. Nanophase and composite materials are immediate candidates for biomimickry. This paper summarizes the main features of biomineralization placing particular emphasis on the importance of specific molecular interactions at interfaces comprising inorganic and organic surfaces. Adaptation of these ideas to synthetic systems involving the integration of supramolecular chemistry (vesicles, micelles, Langmuir monolayers) into materials science is described. Geometric, electrostatic and stereochemical factors are responsible for crystal chemical specificity in these model systems.

1.0 INTRODUCTION

The successful fabrication and utilization of advanced inorganic materials depend on the integration of two key objectives. First, developments are required in the synthetic capabilities of generating new structures tuned to electronic, optical, and magnetic properties. This is the realm of synthetic chemistry and while much progress has been made in organic (polymer) chemistry, achievements in the rational preparation of inorganic materials remain limited. The use of organometallic compounds as tailored precursors to controlled synthesis is gaining momentum and appears promising. Related to the need to produce novel materials is a requirement for more advanced molecular engineering of currently technologically useful products. In principle, the functional use of many materials depends on ultrastructural properties such as crystal size, shape, texture, and organization. For example, catalytic, optical and magnetic properties are often critically dependent on particle shape and size while mechanical strength and toughness are closely related to packing and structural organization.

It is this second requirement that relates closely to biomineralization.[1] In general, biomineralization can be defined as the biological synthesis of inorganic minerals for functional application (Table-I). The structures of these materials, from a solid-state chemical perspective, are relatively simple. There are, for example, no complex phases comparable to high-temperature ceramics or superconductors. But the fabrication of these simple materials is astonishing. For instance, the mineral calcite ($CaCO_3$) can be engineered for use in mechanical support (shells), gravity devices (otoconia), optical lenses (trilobites), and protection (coccoliths). In each case the biomineral is replicated in a precise form commensurate with the functional requirements of a larger scale structure. The mechanisms by which this is achieved and the possibilities of mimicking such processes are the central features of this chapter. We begin with a brief overview of the general relevance of biomineralization to materials science, followed by a discussion of the molecular aspects of controlled biomineralization at inorganic-organic interfaces. In the final section of the paper we describe some recent work on the modeling of biomineralization as a new approach to the crystal engineering of inorganic materials.

2.0 BIOMINERALIZATION AND MATERIALS SCIENCE

The single-celled marine alga *Emiliania huxleyi* is surrounded by a close-packed arrangement of calcite ($CaCO_3$) scales (coccoliths) organized along the outer cell wall (Figure 1(a)). Each coccolith comprises a radial array of complex-shaped units that endow the scale with a double-rimmed perimeter (Figure 1(b)). Individual units of the coccolith have a characteristic species-specific shape consisting of a thin, elongated lower and medial element, a semi-cylindrical central element, and a hammer-shaped upper element. Electron diffraction patterns recorded at different regions of individual units indicate that the complex shape is exhibited by a well-ordered single crystal of calcite oriented along specific crystallographic directions (Figure 2 (a)).[2] The fact that the size and patterning of the crystal form is precisely replicated is remarkable.

Table-I. The Types and Functions of the Main Inorganic Solids Found in Biological Systems

Mineral	Formula	Organism/Function
Calcium Carbonate:		
Calcite	$CaCO_3$*	Algae/exoskeletons, Trilobites/eye lens
Aragonite	$CaCO_3$	Fish/gravity device, mollusks/exoskeleton
Vaterite	$CaCO_3$	Ascidans/spicules
Amorphous	$CaCO_3 \cdot nH_2O$	Plants/Ca store
Calcium Phosphate:		
Hydroxyapatite	$Ca_{10}(PO_4)_6(OH)_2$	Vertebrates/endoskeletons, teeth, Ca store
Octa-calcium phosphate	$Ca_8H_2(PO_4)_6$	Vertebrates/precursor phase in bone?
Amorphous	?	Mussels/Ca store, vertebrates/precursor, phases in bone?
Calcium Oxalate:		
Whewellite	$CaC_2O \cdot H_2O$	Plants/Ca store
Weddellite	$CaC_2O_4 \cdot 2H_2O$	Plants/Ca store
Group 11A Metal Sulfates:		
Gypsum	$CaSO_4$	Jellyfish larvae/gravity device
Barite	$BaSO_4$	Algae/gravity device
Celestite	$SrSO_4$	Acantharia/cellular support
Silicon Dioxide:		
Silica	$SiO_2 \cdot nH_2O$	Algae/exoskeletons
Iron Oxides:		
Magnetite	Fe_3O_4	Bacteria/magnetotaxis chitons/teeth
Goethite	a-FeOOH	Limpets/teeth
Lepidocrocite	g-FeOOH	Chitons (Mollusca)/teeth
Ferrihydrite	$5Fe_2O_3 \cdot 9H_2O$	Animals and plants/Fe storage proteins

* A range of magnesium-substituted calcites are also formed.

Moreover, the alignment and assembly of the coccolith units take place within the cell at the nucleation stage of each crystal (Figure 2(b)). The immature crystals are oriented along the same directions as the mature crystals (J. M. Didymus, unpublished observations). The morphology at the early stages of growth is tabular with no curved edges, suggesting that the factors responsible for oriented nucleation and coccolith shape are distinct and separated in time and space.

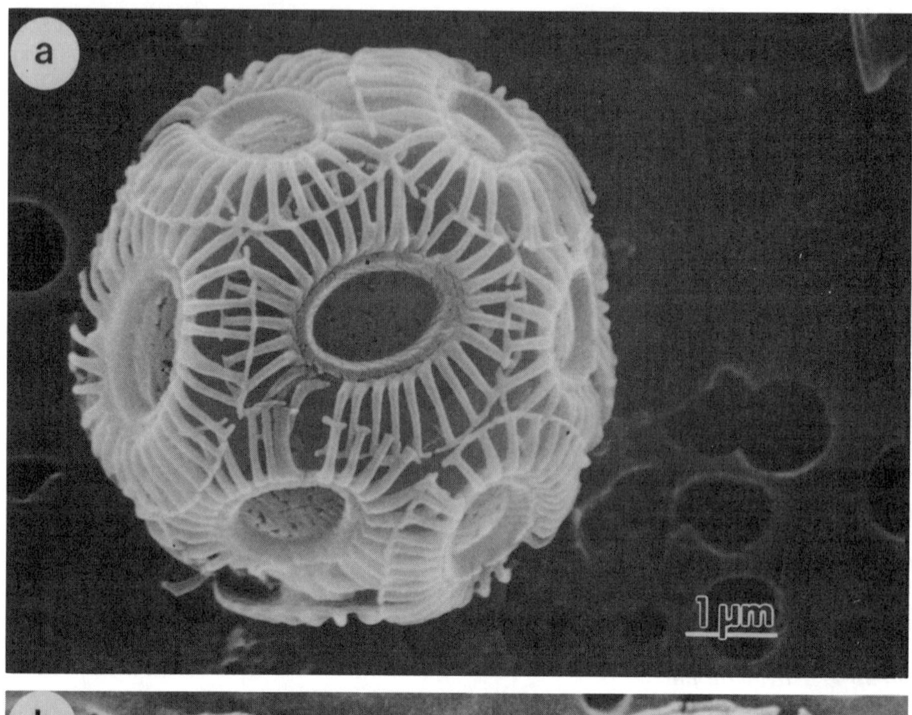

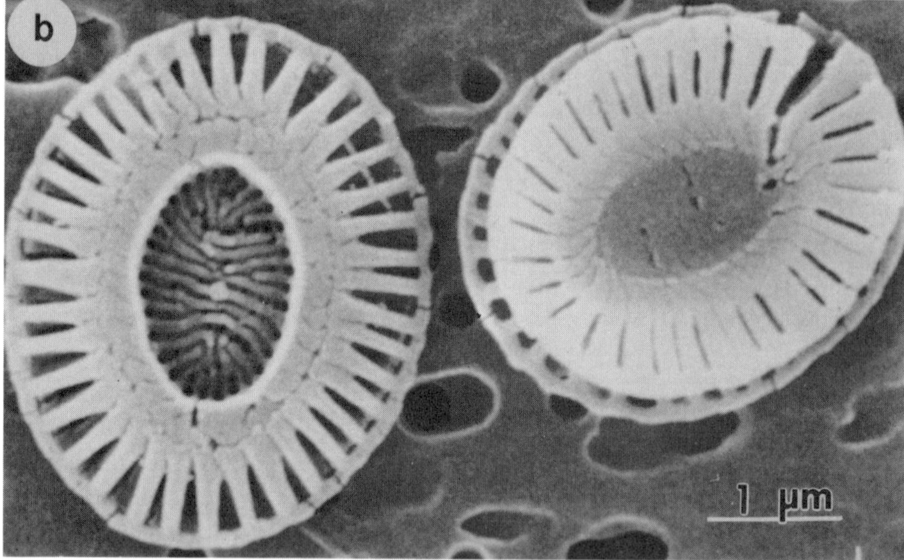

Figure 1. Scanning electron micrographs of (a) whole cell of the alga *Emiliania huxleyi* showing intact coccolithosphere of calcite scales; (b) individual coccoliths viewed distal and proximal to the cell wall. (Photographs courtesy of Dr. J. Young, Natural History Museum, UK.)

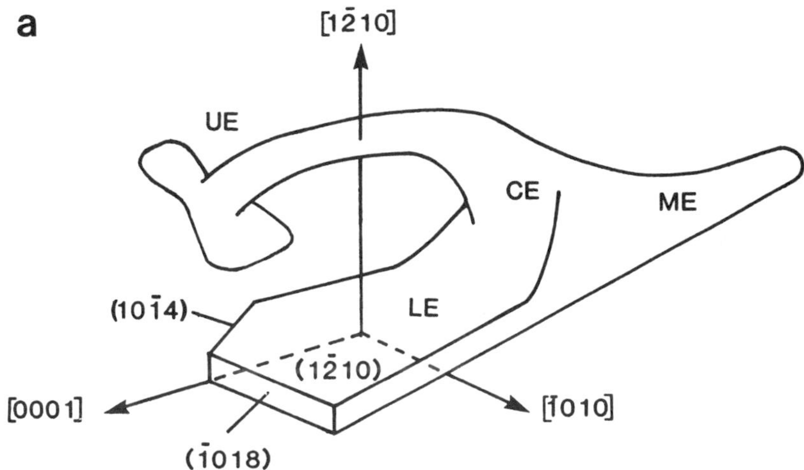

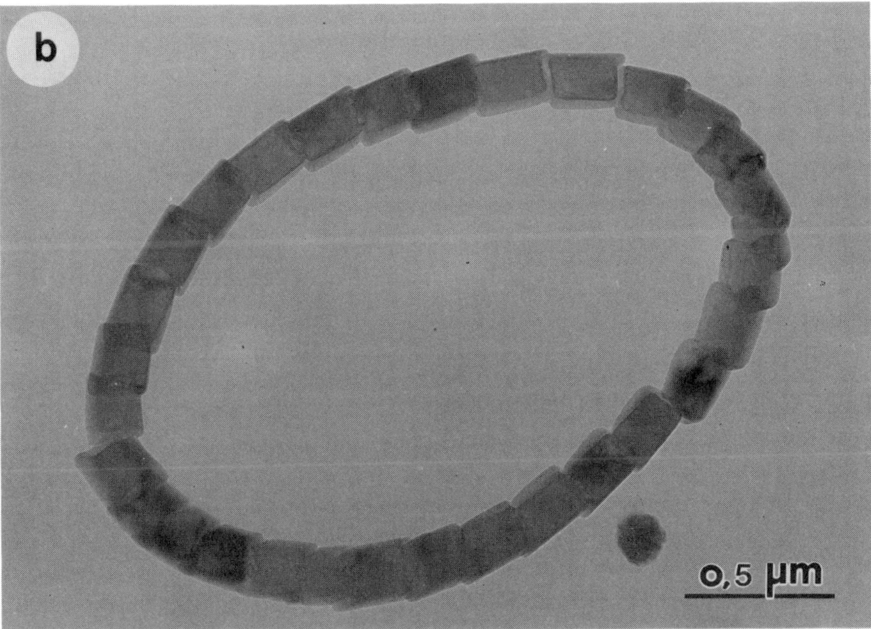

Figure 2. (a) Drawing of a mature coccolith unit showing the unique shape and associated crystallographic directions of a single crystal of calcite. (b) Transmission electron micrograph of a coccolith at early stages of development. (Photograph courtesy of Dr. J. M. Didymus, University of Bath, UK.)

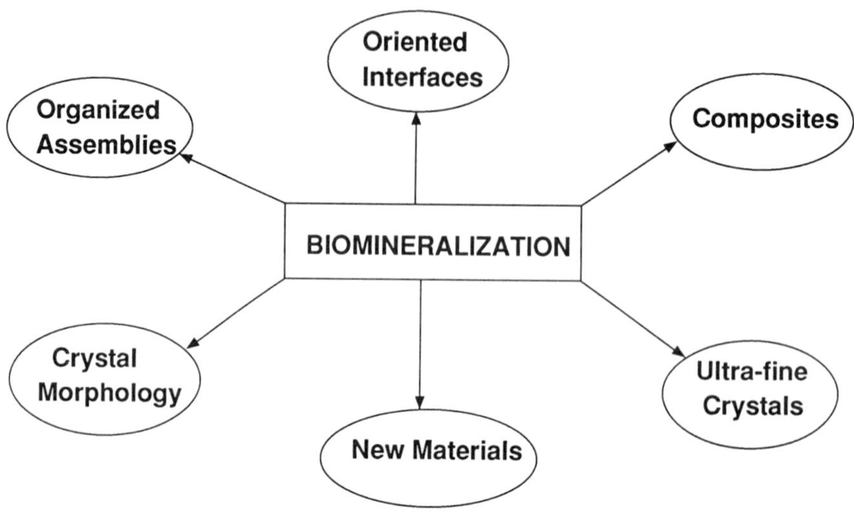

Figure 3. The relationships between biomineralization and materials science.

It is clear from the example in Figure 2 that the control of crystal structure, size, shape, orientation, and assembly are fundamental aspects of biomineralization. Similar objectives can be found in many branches of materials science; Figure 3 summarizes some of the possible areas of overlap. At the simplest level, there are two potential connections between the natural processes of biomineralization and the synthetic demands of materials science:

(i) commercial exploitation of biologically-derived, tailored inorganic materials, and

(ii) assimilation and adaptation of biological concepts and mechanisms into materials design.

The former is an extension of biotechnology in which microbial systems could be utilized in producing mineral powders. Some of the possible processes have been discussed elsewhere.[3] In general, the use of biological sources is only applicable where the high production costs are offset by a marketable speciality product. While this is feasible for organic-based products such as polyhydroxybutyrate, it places a severe limitation when we transfer the approach to biomineralization. Currently, the most potentially viable system is the production of magnetic powders by the large scale cultivation of magnetotactic bacteria.[4] Other possibilities include the use of size- and shape-specific biogenic silicas in chromatography columns, catalyst support beds, or as high-purity precursors to silicon-based materials (e.g., rice husks in the production of silicon carbide fibers[5]). Algal calcareous exoskeletons (corals) have also been used in animal bone implants[6] where the high porosity of these materials aids the infiltration of the implant by surrounding bone tissue.

The adaptation of biomineralization concepts to materials design is currently attracting much attention.[7,8] In essence, the procedure is to simulate the key aspects of molecular control outside the biological context. This provides both a testing of ideas and concepts central to biomineralization and an assessment of their generic importance. The areas of immediate relevance include:

(i) nanoscale synthesis,
(ii) morphological specificity,
(iii) interfacial control of molecular structure and orientation,
(iv) ultrastructural assembly, and
(v) hierarchical composite structures.

Many biological systems exhibit some, if not all, of these features in their synthesis of inorganic minerals. The formation of ultrafine materials, constrained to dimensions on the nanometer scale, is an effective strategy in the storage of available Fe in the protein, ferritin (see below), Cd detoxification (as CdS) in yeasts[9] and in optimizing the magnetic properties of magnetite crystals generated by magnetotactic bacteria.[10] Many biominerals, as illustrated in Figure 1, exhibit unusual morphologies not based on crystallographic symmetry. Others show crystallographic faces but with a reduction in the crystal symmetry. For example, magnetite crystals formed in magnetotactic bacteria are elongated even though they are indexed with a cubic (isometric) symmetry.[11] Crystalline biominerals are often oriented along specific axes and this can result in oriented assemblies such as in coccolith plates and nacre. The nature of these motifs bears a close relationship with the spatial patterning of organic polymers such that complex, multilayered composites (e.g., bones, shells, teeth) are constructed at the microscopic and macroscopic level. The resulting composites often exhibit unique mechanical

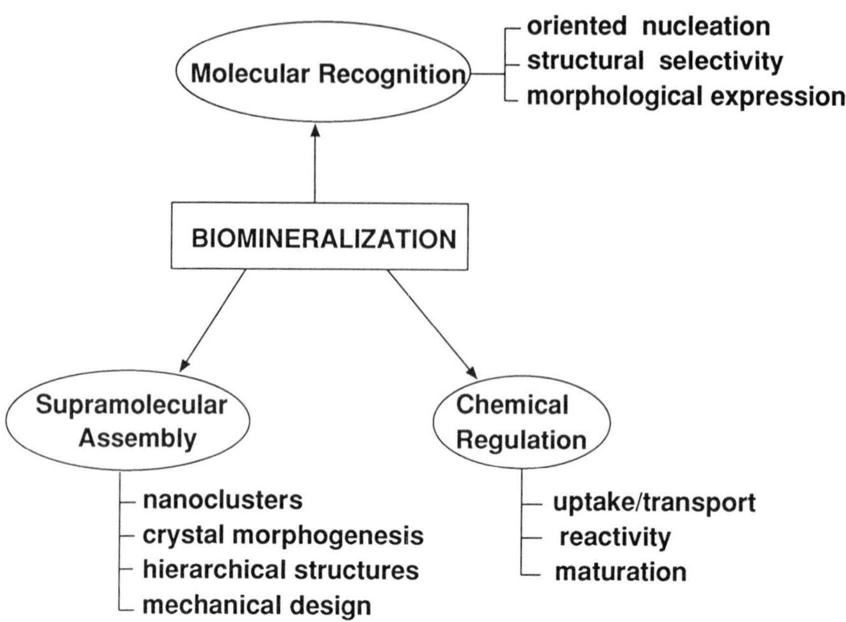

Figure 4. Control factors in biomineralization.

properties and the study of these materials is of profound importance in materials science.

3.0 CONTROL MECHANISMS

The formation of a complex material, such as a coccolith scale, involves many features of control at the site of mineralization. The site must be delineated from the surrounding biological environments, activated at specific times in the life of the organism, constrained in size and shape, and highly regulated with respect to the chemistry of the mineralization process. The principal regulatory factors controlling nucleation and crystal growth are (Figure 4):

(i) spatial delineation by supramolecular assemblies,

(ii) chemical regulation by transport processes, and

(iii) molecular recognition at inorganic-organic interfaces.

The confinement and physical shaping of the mineralization zone by membrane and polymer assemblies is a general feature of most systems involving controlled biomineralization. Entrapment of mineral particles within bilayer vesicles is commonplace. This results in crystals ranging in diameter from 20 nm to several microns. Particles smaller than this lower limit require nanoscale reaction volumes and these can be readily established through the use of polypeptide micelles (Fe oxides in ferritin, CdS in γ-glutamyl complexes in yeasts) and polymeric crystals such as collagen where the mineral is a guest phase housed within interstitial sites of the host lattice. Shaping of these supramolecular systems by stress fields generated by cellular forces can then lead to novel morphologies simply on the basis that the mineral grows into the space made available to it. Superordering of the primary assemblies results in hierarchical structures and associated mechanical and organizational properties.

Details of the processes of chemical regulation of biomineralization are given elsewhere.[12,13] In general, the chemical processes are regulated by transport and/or reaction-mediated mechanisms. The localization of the mineralization event within supramolecular systems enables the supersaturation levels to be precisely regulated through facilitated ion-flux, complexation/decomplexation switches, local redox and pH modifications and changes in local ion activities via ionic strength and vectorial water fluxes. Reaction mechanisms can also influence the kinetics of surface-mediated processes. For example, the stability of embryonic clusters in nucleation, expression of crystal faces, pathways of phase transformations, and aggregation (maturation) are all critically dependent on the relative activation energy barriers of interfacial processes.

It has long been recognized that one of the pivotal issues which separates biomineralization from mineralization is the involvement of macromolecular substrates.[14] In these "organic-matrix-mediated" processes the mineral is often oriented along specific crystallographic directions with respect to an underlying organic surface and the generally accepted hypothesis is that the matrix is responsible for oriented nucleation. In such cases, there must be a degree of complementarity between the structure and surface chemistry of the organic surface and that of the nucleated crystal face. This has lead to the concept of molecular recognition at the interface between inorganic and organic surfaces.[15] Similar interactions exist between soluble macromolecules and crystal surfaces. Although these are relatively unexplored areas compared with recognition processes involving molecules in solution, they are of paramount importance not only in biomineralization but in the development of new approaches to crystal engineering in materials science. We discuss some of the main aspects of recognition at inorganic-organic interfaces in the next section.

4.0 MOLECULAR RECOGNITION AT INORGANIC-ORGANIC INTERFACES

The specific interaction of molecules in solution as exemplified by host-guest complexes of macrocyclic ligands, enzyme-substrate reactions, and antibody-antigen coupling is determined by complementarity in size, charge, and molecular shape, often generated by the interplay of structural, stereochemical and dynamical relationships. Although these factors may be involved in molecular interactions at surfaces, the analogy with solution chemistry is not direct. For example, we can consider the role of an organic matrix to that of an enzyme in solution with the incipient nuclei acting as the corresponding substrate. However, the long-range eletrostatic forces of inorganic surfaces and the requirement of space symmetry imply that other considerations need to be taken into account. Moreover, as a description of the molecular forces operating at idealized inorganic-organic interfaces is not currently available and in no cases do we have knowledge of the structure (periodic, amorphous, polyhedral) size, or composition of critical clusters of whether the initial interactions involve ion-binding or larger scale polynuclear events, it is necessary at this stage to consider recognition in the broadest terms. If we postulate that the role of the matrix is primarily to lower the activation energy of nucleation by decreasing the time between ionic collisions, then this can be achieved by simple catalysis involving complementarity in electrostatic binding. Alternatively, a more complex situation can be envisaged in which the matrix stabilizes a specific conformation of the transition state (nucleus) by structural and stereochemical recognition. Both mechanisms can result in oriented nucleation.

We know several general features about the matrix (Table-II):[16]

(i) *Primary Structures.* The matrix is matched to the coordination chemistry of the ionic species comprising the mineral through the choice of ligands exposed at the interface. For example, calcified invertebrate tissues contain macromolecules rich in carboxylate residues[17] such that Ca binding mimics Ca-CO_3 interactions in the crystal structures of calcite and aragonite. Similarly, Ca-PO_4 lattices can be established through $ROPO_3$-Ca binding

Table-II. Organic Matrices in Biomineralization

Primary Structures	
High Asp:	$CaCO_3$ shells, etc.
High R-OPO_3:	CaP bones, teeth
High Ser/Thr:	Ice bacteria, fish
Secondary/Tertiary Structures	
Planar:	β-sheets
	Stressed membranes
	S-layers
Curved:	Vesicles
	α-helices
	Triple-helices
	Grooves/pockets
Quarternary Structures	
Self-assembly:	24-mer. ferritin
Membrane aggregation:	INA bacteria
Templates:	Shell proteins?

INA = Ice Nucleation Active.

(phosphoproteins in bone[18] and dentine,[19] proteolipids in calcifying bacteria[20]), and silica deposition in diatoms[21] and ice nucleation in bacteria[22] are associated with hydroxy-rich (serine) macromolecules.

(ii) *Secondary and Tertiary Structures.* There are only two options; nucleation at a planar organic surface or at one which is curved. The former can be generated by anti-parallel β-pleated sheets and there is X-ray diffraction evidence[23] to support the role of such structures in biomineralization. Other planar surfaces could be derived by elongation of phospholipid membranes or by crystallization such as the proteinaceous S-layer of bacterial membranes. Curved surfaces are more common; localized protein pockets and grooves, membrane-bound vesicles, α-helical (antifreeze proteins), and triple-helical (collagen) conformations are all possible.

(iii) *Quarternary Structures.* Quarternary structures may play a fundamental role in biomineralization. In evolutionary terms, ferritin (bas bacterioferritin) is an ancient matrix and functions primarily through its ability to self-assemble into a 24-mer capable of sequestering and depositing Fe(III) oxide.[24] Similarly, it is known that the extent of aggregation of membrane-bound proteins is crucial to the efficacy of ice-nucleating bacteria.[25] It seems feasible that the "two-component" model of shell mineralization may also rely on the ordered assembly of the nucleator macromolecules on the relatively inert framework proteins. For example, the EDTA soluble macromolecules of shells are considered to adopt the β-sheet conformation of the underlying structural proteins.[26]

In summary, there are two key structural factors in the use of organic matrices in controlled nucleation. Firstly, the matrix is preorganized with respect to nucleation through processes such as self-assembly, aggregation, membrane vesiculation, and controlled polymerization (cross-linking) which impart spatial regulation of surface functional groups. Second, nucleation at the matrix surface is regiospecific with a limited number of sites being confined to discrete loci. These two factors may be linked in time since ion binding can result in specific conformational changes in the matrix so that nucleation is then initiated within localized domains of the activated structure.

4.1 SITE REACTIVITY AND SPATIAL CHARGE DISTRIBUTION

The matching of charge and polarity distributions across inorganic and organic surfaces is a fundamental aspect of recognition involved in substrate-directed nucleation. It is feasible that the earliest biological approach to regulate nucleation was based on the clustering of charged centers, particularly if these sites were also redox active. The localization of Mn-oxidizing proteins in the cell walls of bacteria such as *Bacillus*[27] and Fe-oxidation centers in ferritin[28] are typical examples. The primary role of charge matching is to favor electrostatic accumulation and hence increase the association time of ions in embryonic clusters. This effect is obviously enhanced if the site is also chemically reactive such that constituents of the nuclei are generated *in situ*. Although no transfer of positional information need take place in this interaction, the topography of the matrix surface may be important. For example, localized pockets or grooves can give rise to high spatial charge densities over dimensions commensurate with stable nuclei (1-5 nm).[13]

The general importance of the above considerations is currently unknown because of the lack of information concerning the primary sequences and tertiary distribution of amino acids in matrix proteins. The best characterized matrix is the iron storage protein, ferritin, for which the crystal structure has been determined at high resolution.[29] The native protein consists of a hollow spherical shell of 24 polypeptide subunits surrounding an internal core (ca. 5 nm in diameter) of the mineral ferrihydrite ($Fe_2O_3 \cdot nH_2O$) [Figures 5(a) and (b)].[29] Ions and molecules can diffuse in and out of the cavity through molecular channels formed at the junction between neighboring subunits.

The ferritin molecule provides an excellent system for investigation because the native iron oxide cores can be removed by reducive dissolution and the resulting intact empty protein shells (apoferritin) can then be remineralized under controlled laboratory conditions. The procedure is simple: incubation of a protein solution with aqueous Fe(II) at pH 7 results in a red-brown coloration due to aerial oxidation and formation of Fe(III) oxide. Amazingly, no bulk precipitation occurs because the resulting mineral is deposited specifically within the protein cavities where it is effectively solubilized. As there is no evidence that the iron oxide cores are oriented with respect to the surrounding protein shell, the ability of ferritin to generate mineral particles is a catalytic process in which oxidation and nucleation proceed at faster rates inside the protein cavity than in the external bulk solution. In the past, assessing the role of the protein in mineralization has been difficult because the subunits are of two different types (so-called H- and L-chains), which are present in variable amounts in the molecule. This has been overcome by the recent success in expressing recombinant ferritins containing 24 subunits of a single type.[30] Moreover, the ability to construct mutant homopolymer ferritins containing site-directed modifications in their amino acids,[31] is an exciting development since, for the first time, a matrix involved in biomineralization can be probed at the molecular level [Figure 5(c)].

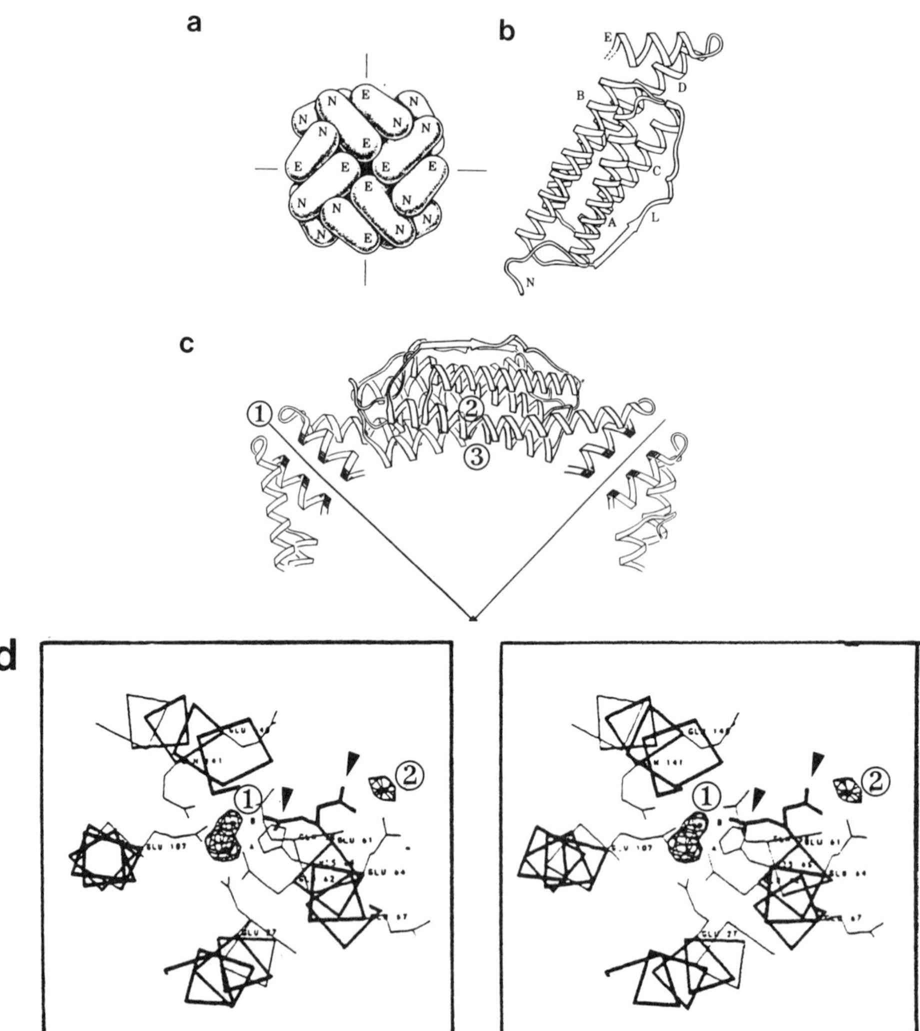

Figure 5. (a) Schematic representation of the ferritin molecule viewed down a fourfold symmetry axis. The iron core resides within the protein shell. Each polypeptide subunit is characterized by a N-terminus (N) and E-helix end region. (b) Ribbon diagram of the α-carbon backbone of an individual subunit showing the 4-helix bundle and a short helix E aligned at an acute angle to the bundle axis. The antiparallel β-sheet loop L is on the outer surface of the protein shell. Residues on the helices B and D constitute the inner cavity wall. (c) View of the apoferritin shell showing approximate locations of sites modified by genetic engineering: (1) along channels, (2) ferroxidase site, (3) inner surface glutamates. (d) Stereo diagram of a 4-helix bundle viewed end on showing the metal binding sites at the ferroxidase center (1) and the inner surface glutamate residues (2). Note that Glu 61 has two alternative positions (arrows). Figures (a) and (b) reproduced with permission from Ref. 29; Figure (c) modified from Ref. 29; and Figure (d) modified from Ref. 28.

The first striking result is that there is a specific oxidation site in the H-chain but not the L-chain polypeptides.[28] The site is partially buried in the 4-helix bundle of the subunit and consists of two neighboring glutamates (Glu 27 and 62) and a histidine (His 65) residue. Interestingly, the ferroxidase center is only 7 Å from a putative nucleation site comprising a cluster of conserved glutamate residues on the inner surface. Modification of the ferroxidase site (Glu 62, His 65 → Lys, Gly) or inner surface glutamates (Glu 61, 64, 67 → Ala) resulted in a reduced rate of mineralization in reconstitution experiments; but, iron cores were formed.[32] By contrast, modification of *both* these sites within the same mutant gave complete inhibition of iron uptake and deposition.[32] The implication is that two sites in this protein act cooperatively in achieving controlled mineralization. When either one of the sites is depleted, the protein remains functional but only at a reduced level. Communication between the ferroxidase site and inner surface glutamates is made possible because residue Glu 61 can adopt two positions: either pendant from the inner surface or as an additional ligand in the coordination sphere of a metal ion bound at the ferroxidase center[28] [Figure 5(d)]. Thus Fe(II) binds at the oxidation site and the resulting Fe(III) ions can be transferred to the inner surface gluta-mates by a pathway involving residue Glu 61. Nucleation of the oxidized species is then facilitated by the high spatial charge density of the inner surface glutamates which reduces the interfacial energy of cluster formation.

These new results highlight the interaction between ligand functionality and quaternary structure. Mineralization can be induced by curved surfaces of appropriate charge density provided that the supply of ions to these regions is effectively controlled. The process will only result in oriented nucleation if specific unicharged crystal faces are preferentially stabilized. Periodicity in the binding sites requires some degree of planarity if long-range structural matches are to be established. This leads to the concept of epitaxy which is discussed in the next section.

4.2 STRUCTURAL AND STEREOCHEMICAL MATCHING

It is important to make a distinction between inorganic epitaxy and "biological epitaxy." The former is well documented and in many cases, geometric matching of lattice parameters is apparent. The situation, however, is not straightforward since large degrees of mismatch can occur and ordered aggregation of nonoriented nuclei to give oriented overgrowths has also been observed at the postnucleation stage. Biological substrates differ from analogous inorganic templates in that they do not show the molecular smoothness or rigidity implicit in epitaxial mechanisms. On the other hand, they exhibit surface stereochemistry due to exposed functional groups and this may be advantageous since it provides for variability in surface structure depending on the binding/ionization state of the residues. Thus, there may be several potential geometric matches for a given crystal face. Alternatively, it is possible that geometric epitaxy is superseded by stereochemical complementarity at the inorganic-organic interface. This can relegate the need for matrix periodicity and is applicable to both planar and nonplanar surfaces. A corollary of a stereochemically driven mechanism is that it may select general rather than specific features of a crystal face. This is important because the putative crystal faces of nuclei are likely to be extensively reconstructed, non-stoichiometric and high in defect sites and general features much as the orientation of carbonate groups may be the only discernable property. In crystallographic terms, nucleation on organic surfaces under such conditions may be selective for specific zone axes rather than particular crystal faces.

The evidence for structural and stereochemical recognition between ionic crystal faces and organic surfaces during nucleation arises primarily from studies using model systems.[33-39] For example, crystallization of calcium carbonate under compressed Langmuir monolayers of stearic acid ($CH_3(CH_2)_{16}COOH$) results in the oriented nucleation of calcite which can be rationalized in terms of charge, stereochemical, and geometric complementarity at the inorganic-organic interface.[38, 39] The crystals are oriented with the {110} face parallel to the organic surface. Prior to nucleation, Ca ions are bound to the negatively charged carboxylate headgroups in a pseudo-hexagonal lattice ($a \approx 5$ Å) determined by the close packing of the surfactant molecules. The interheadgroup spacings are close to those between coplanar Ca atoms in the {110} face of calcite, suggesting a possible epitaxial match. However, a geometric relationship is not the only factor responsible for the nucleation of the {110} face because there are other faces, e.g., (001), where the matching is much closer. A significant difference between the {110} and (001) faces is the orientation of the carbonate anions. They lie perpendicular to the {110} surface but parallel to (001). The stereochemistry of the carboxylate headgroups, therefore, mimics that of the anions in the {110} crystal face but not (001)

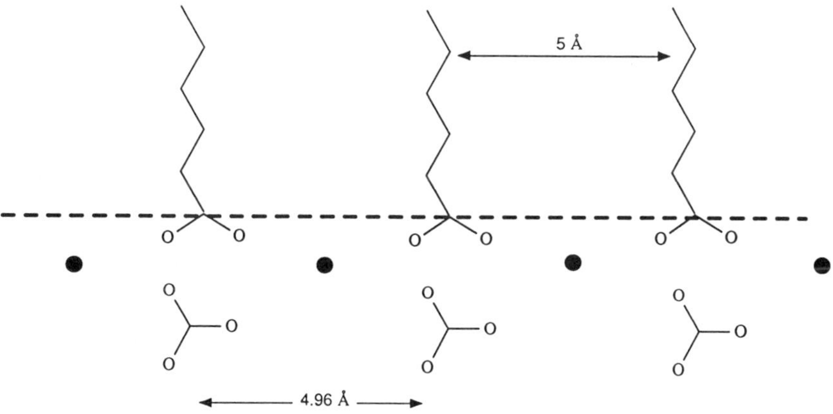

Figure 6. Possible structural and stereochemical relationships between the calcite (110) face and a compressed stearate monolayer. Small filled circles are Ca ions in the Stern layer. Note the stereochemical equivalence of the carboxylate headgroups and the carbonates of the incipient nucleus.

and this can override the reduced geometric matching of the former at the carboxylate surface (Figure 6).

In summary, electrostatic, geometric, and stereochemical recognition processes can occur at mineral-matrix interfaces. Understanding these interactions depends very much on determining the nature of assembled organic surfaces and in elucidating the surface structure of inorganic clusters. Although we have discussed these factors with regard to nucleation, they are also applicable to crystal growth, particularly with respect to the influence of molecular additives on crystal morphology. In these systems, the situation is somewhat reversed since the crystal surface represents the immobilized phase while the organic constituents are dynamic. But in both cases, the adaptation of these concepts to synthetic systems is relevant in the crystal engineering of tailored inorganic materials. Some recent developments in this area are described in section 5.

5.0 CRYSTAL ENGINEERING

There are several possible approaches to the mediation of inorganic materials synthesis by biomimickry:

(i) use of supramolecular assemblies in nanoscale synthesis,
(ii) organic surfaces as molecular templates for nucleation,
(iii) organic additives in the control of crystal morphology, and
(iv) synthetic polymeric matrices as frameworks for composite structures.

Unilamellar vesicles can be readily prepared by ultrasonication of aqueous dispersions of phospholipids. The vesicles have a bilayer membrane enclosing an internal aqueous volume often only 20-50 nm in diameter. When formed in the presence of metal ions, the internal space contains encapsulated species which can subsequently undergo *in situ* precipitation reactions with membrane-permeable species such as OH^- and H_2S [Figure 7(a)]. In effect, this is solid-state chemistry in a very small soap bubble. Materials such as Ag_2O,[40] $CaSiO_3$,[41] Ca phosphates,[42] $Fe_2O_3{\cdot}nH_2O$ and Fe_3O_4,[43, 44] Al_2O_3,[45] and $BaHPO_4$ and SiO_2[46] have been prepared as nanometer-size intravesicular particles. The curved organic membrane has a significant influence on these crystallization reactions. For example, changes in crystal structure, size, and morphology are observed compared with analogous reactions undertaken in the absence of the vesicles.

Recently, we have shown that a similar approach can be applied to the synthesis of nanophase materials using the protein shell of ferritin[47] [Figure 7(b)]. Iron sulfide cores are rapidly synthesized by *in situ* chemical reaction of the native iron oxide cores with H_2S gas. A second approach is to use metal-ion systems that mirror the Fe(II) oxidation and nucleation processes responsible for the specific reconstitution of Fe(III) oxide cores in apoferritin (see section 4.1). An obvious candidate is Mn(II). Indeed, incubation of the empty protein cages with $MnCl_2$ at a pH of 9 results in specific ion uptake, oxidation, and precipitation of amorphous Mn oxide. A third approach is to use hydrolytic polymerization reactions involving cations such as UO_2^{2+} as a basis for inorganic precipitation within the apoferritin cavity.

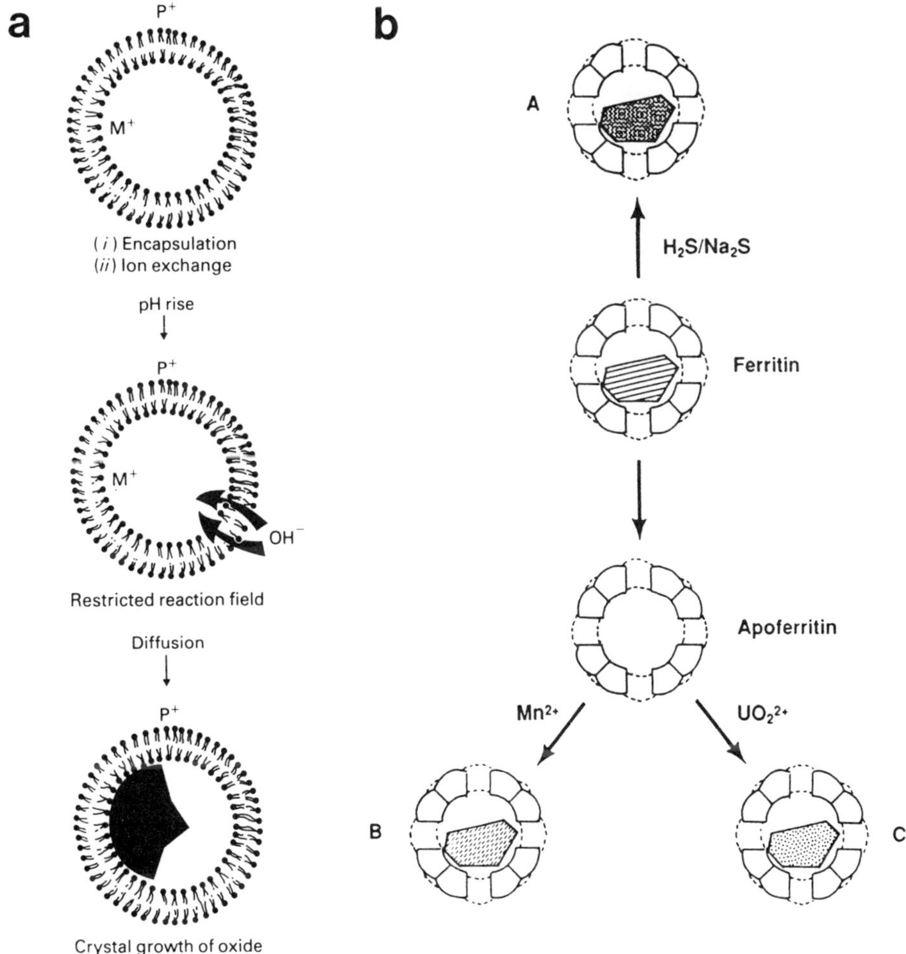

Figure 7. The use of supramolecular assemblies in the nanoscale synthesis of inorganic materials. (a) Oxide formation in phospholipid vesicles, and (b) Fe sulphide (A), Mn oxide (B) and uranyl oxide deposition in ferritin.

The use of organic surfaces as molecular templates in oriented nucleation has attracted much attention in recent years.[48] Earlier work showed that the number of oriented calcite crystals nucleated on a glass surface increased significantly in the presence of adsorbed mollusk shell proteins.[34] Recent investigations have used compressed monolayer films of surfactant molecules formed at the air/water interface as organized templates for inorganic nucleation. The influence of charged or polar headgroups on the oriented nucleation of amino acids and NaCl,[36, 38] and $CaCO_3$ [37-39, 49] have been reported. Studies on the crystallization of $SrSO_4$ on Langmuir-Blodgett multilayer assemblies have also been undertaken.[50]

The influence of these monolayers is very striking. For example, calcite nucleation in the absence of a monolayer results in a random mass of rhombohedral crystals at the solution surface. When a monolayer of stearic acid is present, however, discrete oriented crystals of calcite are nucleated under the organic film (Figure 8). As discussed in Section 4.2, the crystals are oriented due to specific molecular interactions at the interface. Changing the surfactant headgroup from the negatively charged carboxylate to a positively charged amine group results in a dramatic change in the crystal structure. Under these conditions, oriented crystals of the metastable mineral, vaterite, are nucleated.[51, 52]

Another approach is to use tailor-made additives that modify the crystal shape through specific interactions of the adsorbed molecules at the crystal faces. In the past, many additives have been used on an empirical basis and it is only very recently that stereochemical arguments have been developed to explain their effects on inorganic crystallization.[33, 53-55] Many of the soluble macromolecules associated with biominerals are considered to act in this manner. For example, acidic glycoproteins extracted from adult sea urchin tests have been shown to bind specifically to the {110} faces of synthetic calcite crystals.[53] Certain low molecular weight organic acids have a similar effect. For example, calcite crystals grown in aqueous solutions containing variable levels of functionalized and non-functionalized α, ω-dicarboxylates have elongated prismatic morphologies.[54] A similar habit is observed for crystals grown in the presence of inorganic phosphate and, to a lesser extent, sulfate.[55] But the specificity of the additives is very dependent on molecular structure and even apparently minor changes can have a dramatic influence on the interaction at the crystal surface. Elucidation of the structural, electrostatic and stereochemical factors involved in these processes is a necessary requirement if additives are to have general application in tailored materials synthesis.

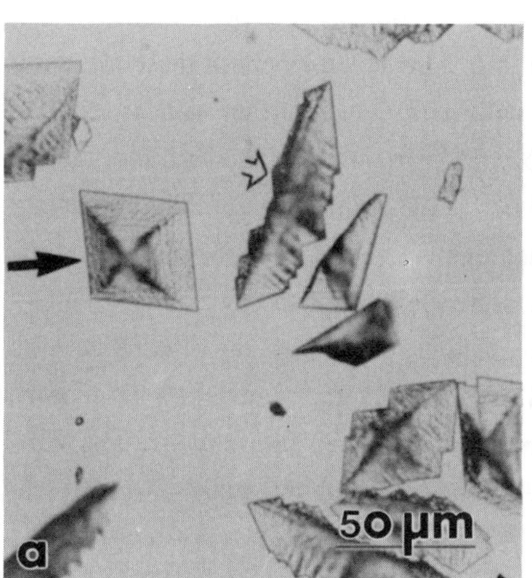

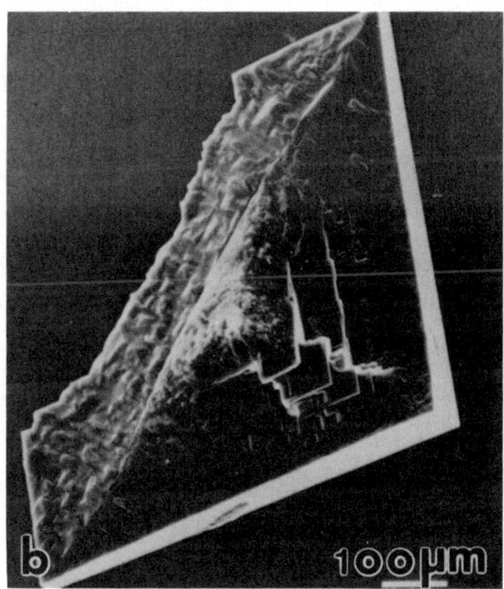

Figure 8. Oriented calcite formed under a compressed stearate monolayer, (a) optical micrograph; (b) scanning electron micrograph of individual crystal.

Finally, the potential for precipitating inorganic materials *in situ* within polymeric matrices is receiving attention as a means of producing highly structured composite materials.[7] Some of the possibilities have been recently reviewed.[56] The major problem lies with the difficulty of attaining high volume fractions of the inorganic phase. Recent studies using partially crystalline polyethylene oxide gels as a structured matrix for CdS crystallization have indicated that oriented nucleation can be achieved in these systems by using surfactant additives.[57] The combination of nucleator molecules and framework polymers is a good analogy to biomineralization and is a promising approach.

6.0 CONCLUSIONS

The modeling of biomineralization is an innovative approach to controlled materials synthesis. In particular, the use of enclosed supramolecular assemblies appears to be a viable low-temperature route to nanophase inorganic materials. A complimentary approach involving planar assemblies of organic macromolecules has potential in the oriented nucleation of organized inorganic materials. The development of these ideas into polymer-based inorganic composites has been initiated. Much further work is required particularly in establishing preferred crystallographic orientations and high inorganic volume fractions. One process yet to be explored is the ability to assemble inorganic materials into organized microarchitectures analogous to bone, teeth, or shells. In biological systems, the assembly of bioinorganic minerals is regulated by flow processes whereby cells control the fluxes of ions and molecules to and from the mineralization sites. Concepts such as self-assembly, dynamics, feedback and remodeling in inorganic materials synthesis are still figments of our imagination. Their future realization will revolutionize the design and fabrication of advanced inorganic materials.

7.0 ACKNOWLEDGEMENTS

I wish to thank Professor P. M. Harrison (Sheffield), Dr. P. Arosio (Milan), V. J. Wade and F. C. Meldrum (Bath) for their involvement in studies of ferritin, and Dr. B. R. Heywood (Salford) and J. M. Didymus (Bath) for their contribution to nucleation and morphological studies of calcite.

8.0 REFERENCES

1. *Biomineralization: Chemical and Biochemical Perspectives,* S. Mann, J. Webb, and R. J. P. Williams (eds.) (VCH Publishers, Weinheim, 1989).

2. S. Mann and N.H. C. Sparks, *Proc. R. Soc. Lond. B*, **234**, 441-453 (1988).

3. S. Mann and P. D. Calvert, *Trends in Biotechnology*, **5**, 309-314 (1987).

4. T. Matsunaga and S. Kamiya, *Appl. Microbiol. Biotechn.*, **26**, 328-333 (1987).

5. N. K. Sharma, W. S. Williams, and A. Zangvil, *J. Am. Ceram. Soc.*, **67**, 715-720 (1984).

6. D. Roy, *New Scientist,* **72**, 163-165 (1976).

7. S. Mann, *Nature*, **349**, 285-286 (1991).

8. *Materials Synthesis Utilizing Biological Processes, Mater. Res. Soc. Symp. Proc.*, Vol. 174, P. C. Reike, P. D. Calvert, and M. Alper (eds.) (Materials Research Society, Pittsburgh, 1990).

9. C. T. Dameron, R. N. Reese, R. K. Mehra, A. R. Kortan, P. J. Carroll, M. L. Steigerwald, L. E. Brus, and D. R. Winge, *Nature,* **338**, 596-597 (1989).

10. R. P. Blakemore and R. F. Frankel, *Scientific American*, **245**, 42-49 (1981).

11. S. Mann, N. H. C. Sparks, and R. P. Blakemore, *Proc. R. Soc. Lond.* B, **231,** 477-487 (1987).

12. S. Mann, *Structure and Bonding*, **54**, 125-174 (1983).

13. S. Mann in *Biomineralisation: Chemical and Biochemical Perspectives*, S. Mann, J. Webb, and R. J. P. Williams (eds.) (VCH Publ., Weinheim, 1989), pp. 35-62.

14. H. A. Lowenstam, *Science*, **211**, 1126-1131 (1981).

15. S. Mann, *Nature*, **332**, 119-124 (1988).

16. S. Mann, B. R. Heywood, S. Rajam, and V. J. Wade, in *Proceedings of the Sixth International Conference on Biomineralization*, S. Suga (ed.) (Springer-Verlag, Berlin, 1991) pp. 47-56.

17. S. Weiner, *CRC Crt. Rev. Biochem.*, **20**, 365-408 (1986).

18. M. J. Glimcher, *Phil. Trans. R. Soc. Lond. B*, **304**, 479-508 (1984).

19. S. L. Lee, A. Veis, and T. Glonek, *Biochemistry*, **16**, 2971-2979 (1977).

20. B. Boyan-Salyers, and A. Boskey, *Calc. Tiss. Intern.,* **30**, 167-174 (1980).

21. D. M. Swift and A. P. Wheeler, in *Surface Reactive Peptides and Polymers*, C. S. Sikes and A. P. Wheeler (eds.) ACS Symp. Series 444 (ACS, 1991) pp. 340-353.

22. P. K. Wolber and G. J. Warren, *Trends Biol. Sci.*, **14**, 179-182 (1989).

23. S. Weiner and W. Traub, *FEBS Letts.,* **111**, 311-316 (1980).

24. P. M. Harrison, P. J. Artymiuk, G. C. Ford, D. M. Lawson, J. M. A. Smith, A. Treffry, and J. L. White, in *Biomineralisation: Chemical and Biochemical Perspectives,* S. Mann, and J. Webb, and R. J. P. Williams (eds.) (VCH Publishers, Weinheim, 1989) pp. 257-294.

25. G. J. Warren and P. K. Wolber, *Cryo-Lett,.* **8**, 204-215 (1987).

26. S. Weiner and W. Traub, *Phil. Trans. R. Soc. Lond. B*, **304**, 425-434 (1984).

27. J. P. M. de Vrind, E. W. de Vrind-de Jong, J. W. de Voogt, P. Westbroek, F. C. Boogerd, and R. A. An Rosson, *Appl. Envir. Microbiol.,* **52**, 1096-1100 (1986).

28. D. M. Lawson, P. J. Artymiuk, S. J. Yewdall, J. M. A. Smith, J. C. Livingstone, A. Treffry, A. Luzzago, S. Levi, and P. Arosio, G. Cesareni, C. D. Thomas, W. V. Shaw, and P. M. Harrison, *Nature,* **349**, 541-544 (1991).

29. G. C. Ford, P. M. Harrison, D. W. Rice, J. M. A. Smith, A. Treffry, J. L. White, and J. Yariv, *Phil. Trans. R. Soc. Lond. B,* **304**, 551-565 (1984).

30. L. Levi, A. Luzzago, G. Cesareni, A. Cozzi, F. Franceschinelli, A. Aobertini, and P. Arosio, *J. Biol. Chem.,* **263**, 18086-18092 (1988).

31. S. Levi, J. Salfeld, F. Franceschinelli, A. Cozzi, M. Dorner, and P. Arosio, *Biochemistry,* **28**, 5179-5184 (1989).

32. V. J. Wade, S. Levi, P. Arosio, A. Treffry, P. M. Harrison, and S. Mann, *J. Mol. Biol.,* in press.

33. L. Addadi and S. Weiner, *Proc. Natl. Acad. Sci. USA,* **82**, 4110-4114 (1985).

34. L. Addadi, J. Moradian, E. Chay, N. G. Maroudas, and S. Weiner, *Proc. Natl. Acad. Sci. USA*, **84**, 2732-2736 (1987).

35. L. Addadi and S. Weiner, in *Biomineralisation: Chemical and Biochemical Perspectives,* edited by S. Mann, J. Webb, and R. J. P. Williams (eds.) (VCH Publishers, Weinheim, 1989) pp. 134-156.

36. E. M. Landau, R. Popovitz-Bior, M. Levanon, L. Leiserowitz, M. Lehav, and J. Sagiv, *Molec. Cryst. Liq. Cryst.,* **134**, 323-335 (1986).

37. S. Mann, B. R. Hewyood, S. Rajam, and J. D. Birchall, *Nature,* **334**, 692-695 (1988).

38. S. Mann, B. R. Heywood, S. Rajam, J. B. A. Walker, R. J. Davey, and J. D. Birchall, *Adv. Materials,* **2**, 257-261 (1990).

39. S. Mann, B. R. Heywood, S. Rajam, and J. B. A. Walker, *J. Appl. Phys.,* **24**, 154-164 (1991).

40. S. Mann and R. J. P. Williams, *J. Chem. Soc. Dalton Trans.,* 311-316 (1986).

41. S. Mann, A. J. Skarnulis, and R. J. P. Williams, *Israel J. Chem.,* **21**, 3-7 (1981).

42. B. R. Heywood and E. D. Eanes, *Calcif. Tissue Int.,* **41**, 192-201 (1987).

43. S. Mann, J. P. Hannington, and R. J. P. Williams, *Nature,* **324**, 565-567 (1986).

44. S. Mann and J. P. Hannington, *J. Colloid Interface Sci.,* **122**, 326-335 (1988).

45. S. Bhandarkar and A. Bose, *J. Colloid Interface Sci.,* **135**, 531 (1990).

46. C. C. Perry, in *Proceedings of the 1st A.N.A.I.C. Conference on Si and Sn* (OUP), in press.

47. F. C. Meldrum, V. J. Wade, D. L. Minno, B. R. Heywood, and S. Mann, *Nature,* **349**, 684-687 (1991).

48. E. M. Landau, S. G. Wolf, M. Levanon, L. Leiserowitz, M. Lehav., and J. Sagiv., *J. Am. Chem. Soc.,* **111**, 1436-1445 (1989).

49. S. Mann, B. R. Heywood, S. Rajam, and J. D. Birchall, *Proc. R. Soc. Lond. A,* **423**, 457-471 (1989).

50. N. P. Hughes, D. Heard, C. C. Perry, and R. J. P. Williams, *J. Appl. Phys.,* in press.

51. S. Rajam, B. R. Heywood, J. B. A. Walker, R. J. Davey, J. D. Birchall, and S. J. Mann, *Chem. Soc. Faraday Trans. 1,* **87**, 727-734 (1991).

52. B. R. Heywood and S. Mann, *J. Chem. Soc. Faraday Trans.,* **87**, 735-743 (1991).

53. A. Berman, L. Addadi, and S. Weiner, *Nature,* **331**, 546-548 (1988).

54. S. Mann, J. M. Didymus, N. P. Sanderson, B. R. Heywood, and E. J. A. Samper, *J Chem. Soc. Faraday Trans.,* **86**, 1873-1880 (1990).

55. J. M. Didymus, S. Mann, N. P. Sanderson, B. R. Heywood, and E. J. A. Samper, in *Proceedings of the Sixth International Conference on Biomineralization,* S. Suga (ed.) (Springer-Verlag, Berlin, 1991) pp. 267-272.

56. P. D. Calvert and S. Mann, *J. Mat. Sci.,* **23**, 3801-3815 (1988).

57. P. A. Bianconi, J. Lin, and A. R. Strzelecki, *Nature,* **349**, 315-317 (1991).

Color Plate 1. The morphology of the collagen fibril in tendon. Viewed between crossed polarizers in the optical microscope, the collagen fibrils that make up the tendon have a wavy appearance. Upon further examination this waveform is characterized as a planar zigzag. Adjacent fibrils are all crimped in register. It is this wavy conformation of the fibrils on the microscopic level of the structure that imparts a high degree of elasitcity to the tendon, enabling it to be stretched repeatedly longitudinally without damaging the underlying structure on the nano- and molecular levels (Ref. 1). (*Black and white version of figure available on page 17.*)

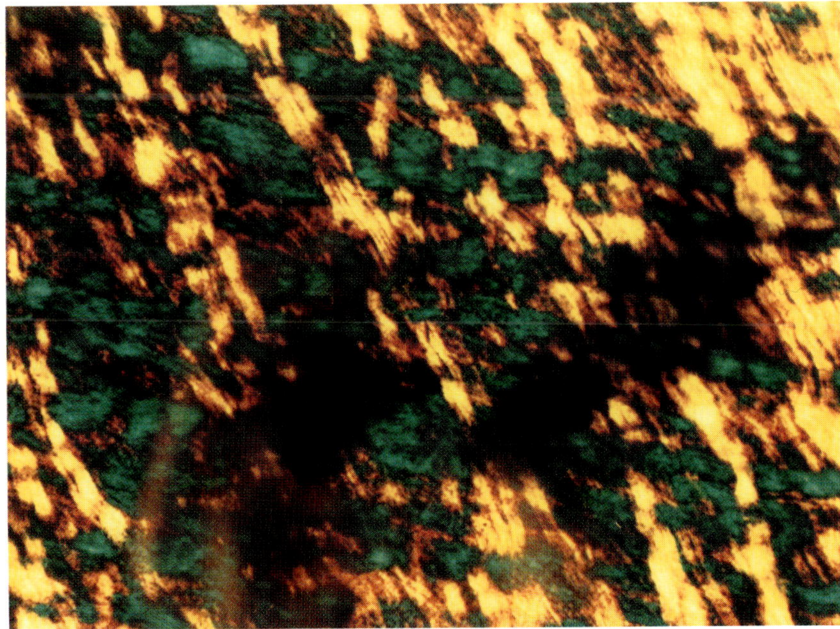

Color Plate 2. The biaxial crimped layered structure of collagen fibers in the intestine. Collagen fibers in the intestine have the same planar crimped morphology observed in the tendon. The parameters of the waveforms (crimp angles, period) are different, however, with the greater extensibility required in the intestine being reflected in a large crimp angle (Ref. 9). (*Black and white version of figure available on page 28.*)

Color Plate 3. OM micrograph of a stained section of *Odontotaenius disjunctus* elytra cuticle exhibits the observed staining gradient. (*Black and white version of figure available on page 170.*)

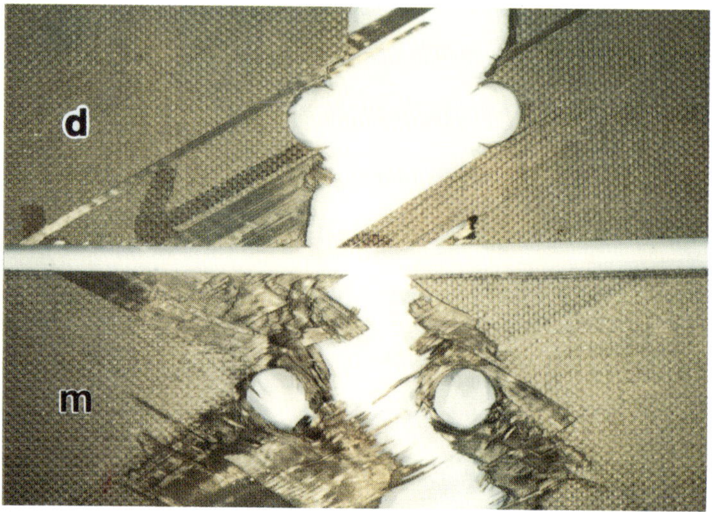

Color Plate 4. A failed $[\pm 30°]_{4S}$ drilled-(d) and molded-(m) hole specimen and their different failure characteristics. (*Black and white version of figure available on page 187.*)

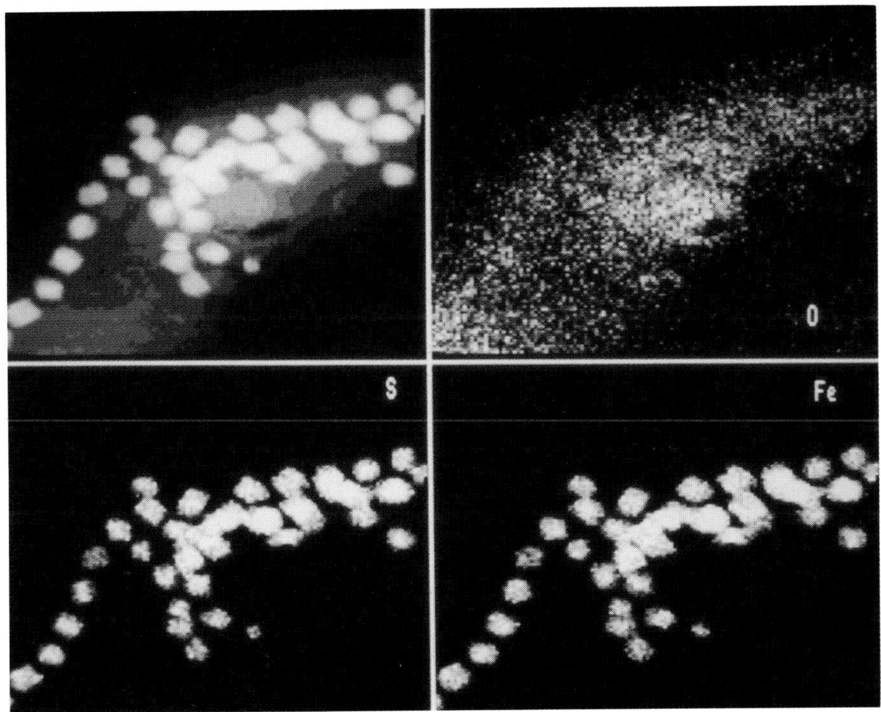

Color Plate 5. Elemental density maps for the particles in one of the constituent cells of the magnetotactic, multicellular prokaryote: upper left, transmission electron micrograph; upper right, O; bottom left, S; bottom right, Fe. Particles consist of greigite, Fe_3S_4, and pyrite, FeS_2 (Ref. 21). (*Black and white version of figure available on page 212.*)

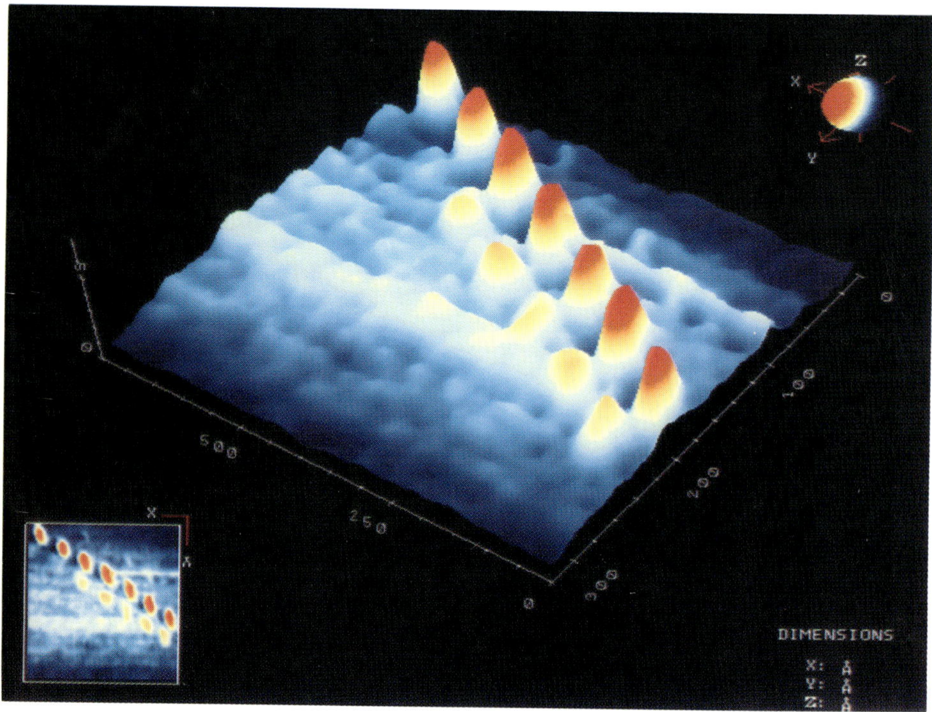

Color Plate 6. A three-dimensional STM image of ZnS particles on HOPG, *in situ* generated on a zinc–arachidate monolayer by 5 min exposure to H_2S. A bird's eye view of the same area in two dimensions (x and y axes are shown on the same scale as in the three-dimensional plot) is shown in the insert. (*Black and white version of figure available on page 232.*)

MICROSTRUCTURE-PROPERTY RELATIONS IN VERTEBRATE BONY HARD TISSUES: MICRODAMAGE AND TOUGHNESS

John D. Currey, Peter Zioupos, and Andy Sedman

Department of Biology, University of York,
York YO1, 5DD, United Kingdom

Bone is a composite composed mainly of collagen, apatite, and water. There is considerable variation in mechanical properties between different bones. Variation in several of these properties can be explained, statistically, by variation in mineral content. A high mineral content is associated with a high Young's modulus, a high yield stress, and a low ultimate strain. High work of fracture is associated with intermediate values of mineral. Yield in bone is not like yield in metals, but is associated with the production of damage, in the form of microcracking. We have examined this microcracking, using optical methods including scanning confocal microscopy, and find that it is much more widespread in materials with a lower mineral content, such as antler, than in more "standard" bone. Fractal analysis of fracture surfaces of various bones, and antler, gives some idea of how the toughness of antler is achieved.

1.0 INTRODUCTION

Vertebrate bony tissues show a wide range of mechanical properties. We know most about tensile strength, stiffness, and work of fracture. Tensile strength ranges from about 20 to 260 MPa, Young's modulus from 5 to 35 GPa, and work of fracture from 20 to 7 kJ m^{-2}. In general, variations in stiffness are determined to a very large extent by the proportion of the tissue that is occupied by the characteristic mineral of vertebrate hard tissues: hydroxyapatite. The porosity of the material also has a considerable effect on its stiffness but, as far as we know, variations in microstructure have much less effect. Tensile strength is also to some extent affected by variations in mineralization, but microstructure is relatively more important. Finally, toughness, as measured by the work of fracture under stable fracture conditions, is related to mineralization, but we consider it probable that the low mineralization allows the microstructure to deflect

Table-I. Mechanical properties of various vertebrate bony tissues. Testing conditions (cross head speed, specimen shape, and so on) were the same for all specimens. The figures are median values of various properties for tensile specimens from various bones from various species. 'MVf': Mineral volume fraction (volume of mineral in parts per thousand); 'E': Young's modulus of elasticity (GPa). (The use of the term 'Young's modulus' implies that there are no strain-rate-dependent effects present. In fact viscoelasticity and damage may tend to reduce the apparent material stiffness somewhat); 'σ_{uit}': Ultimate tensile stress (MPa); ε_{uit}: Ultimate tensile strain; 'W' Work under the stress-strain curve (MJ m). Note: the work under the stress-strain curve may be sensitive to specimen shape and, in brittle materials, to the specimen volume. Therefore, this property should be used only for comparative data, as here, derived from specimens all with the same dimensions.

Species and Tissue	*MVf*	*E*	σ_{uit}	ε_{uit}	*W*
Red deer, immature antler	281	10.0	250	.109	15.6
Red deer, mature antler	287	7.2	158	.114	9.3
Reindeer, antler	300	8.1	95	.051	3.2
Polar bear (3 mo.), femur	328	6.7	85	.044	3.2
Narwhal, tusk cement	331	5.3	84	.060	3.0
Narwhal, tusk dentine	340	10.3	120	.037	3.7
Sarus crane, tarsometatarsus	341	23.1	218	.018	2.0
Walrus, humerus	352	14.2	105	.026	1.4
Fallow deer, radius	360	25.5	213	.019	2.1
Human, femur	362	16.7	166	.029	2.8
Bovine, tibia	364	19.7	146	.018	1.8
Polar bear (9 mo.), femur	366	11.2	137	.042	3.3
Leopard, femur	375	21.5	215	.034	3.4
Brown bear, femur	377	16.9	152	.032	2.3
Donkey, radius	381	15.3	114	.020	1.6
Sarus crane, tibiotarsus	382	23.5	254	.031	4.1
Flamingo, tibiotarsus	382	28.2	212	.013	1.4
Roe deer, femur	383	18.4	150	.011	0.9
Polar bear (3.5 y.), femur	386	18.5	154	.022	2.2
King penguin, radius	394	22.1	195	.010	0.8
Horse, femur	395	24.5	152	.008	0.5
Polar bear (3 y.), femur	397	16.5	142	.028	2.5
Wallaby, tibia	402	25.4	184	.010	1.1
Bovine, femur	410	26.1	148	.004	0.3
Polar bear, femur	414	22.2	161	.020	1.7
King penguin, ulna	421	22.9	193	.011	1.2
Axis deer, femur	428	31.6	221	.019	2.4
Fallow deer, tibia	430	26.8	131	.006	0.4
Wallaby, femur	437	21.8	183	.009	0.8
King penguin, humerus	453	22.8	175	.008	0.7
Fin whale, ear bone	560	34.1	27	.002	0.02

microcracks, and make them innocuous and incapable of growing. Unfortunately we have almost no data on this. The aim of this review is to describe these variations in mineralization and microstructure and to relate them to variations in mechanical properties. In particular, we shall be concerned with toughening mechanisms, and the part that 'controlled' microdamage may or may not play in affecting toughness.

2.0 MECHANICAL PROPERTIES

2.1 YOUNG'S MODULUS

Table-I gives some values of mechanical properties of vertebrate hard tissues, arranged in increasing rank of volume fraction of mineral, all measured under very similar conditions of strain rate, temperature, specimen shape, and so on. For vertebrates, the stiffness of their skeletal structures is usually of overriding importance, and very often toughness is relatively unimportant. Young's modulus is a fairly structure-insensitive mechanical property, and in general there is a good relationship between mineral volume fraction and Young's modulus. For bone, which is essentially a composite of three components: hydroxyapatite, collagen, and water, the value of Young's modulus is determined to quite a high degree by two variables only: calcium content (or mineral volume fraction) and porosity. Different bones are porous to various degrees that depend on the metabolic state of the bone and other factors. These factors probably concern the necessity of destroying bone that has more than its share of microcracks, and of releasing calcium or phosphorus into the blood for homeostatic purposes. Also, when bone is growing rapidly in width the mode of growth requires there to be holes in the bone initially, which are only gradually filled in later.

Figure 1 shows the relationship between Young's modulus and mineral content in a variety of vertebrate bones. Although it is clear, it is not a very clean relationship, R^2 being only 61%. However, if porosity is added as an explanatory variable, the fit becomes much better (Figure 2), R^2 being 75%. That is to say, three quarters of all the variation in Young's modulus can be explained (in statistical terms), by variation in mineral content and porosity.[1]

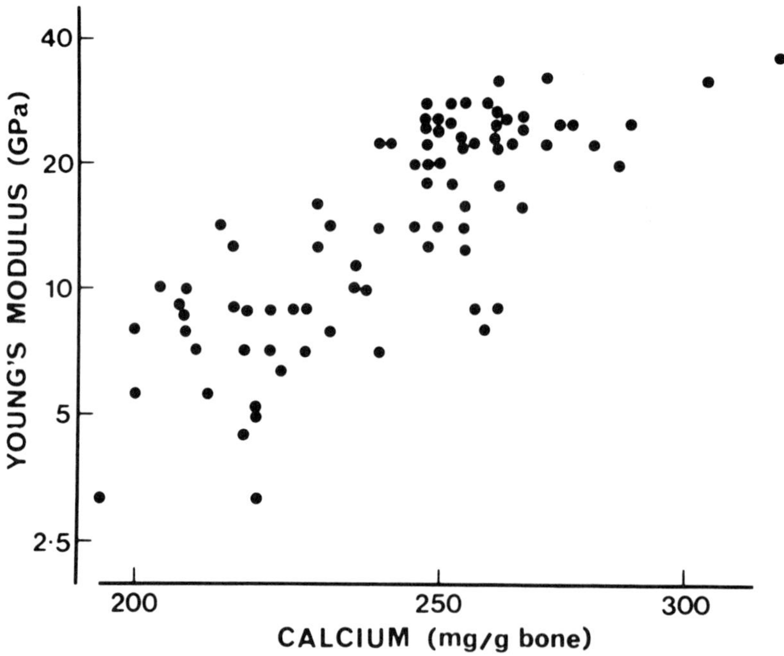

Figure 1. Young's modulus as a function of calcium content. Note log scale on ordinate.

Table-I shows some exceptions; in particular the sarus crane tarsometatarsus and fallow deer radius are anomalously stiff. It is probable that the reason for the high stiffnesses is that the slender bones of the crane and the deer are extremely anisotropic, having a grain very highly oriented along the long axis of the bone.

2.2 TENSILE FAILURE PROPERTIES

Work in our laboratory[2,3] has shown that the compact bone of reptiles, birds, and mammals exhibits considerable variation in strength properties when loaded in tension. Mineral volume fraction, porosity, and orientation can 'explain,' in statistical terms, a great proportion of this variation for some properties, but for many properties it cannot.

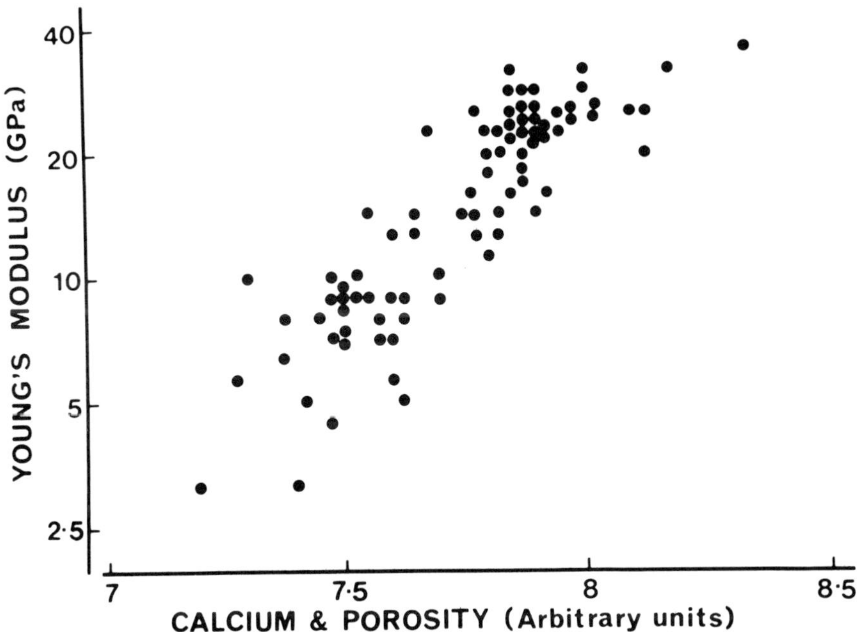

Figure 2. Young's modulus as function of both calcium content and porosity. Note log scale on ordinate.

Perhaps the most striking of the findings are, first, the large amount of post-yield stress and strain shown by the less mineralized bones, particularly antler, and second, the very great variation in the ultimate strain of different bones, and the extent to which this is related to mineralization. Usually, increasing mineralization has a strong negative effect on both the extra stress and strain that are reached after the specimen has yielded.

The effect of mineralization and porosity on various tensile mechanical properties of bone are discussed below.

2.3 YIELD STRESS

Plots of the log of yield stress against mineralization and apparent density (1 - porosity) are shown in Figures 3 and 4. No obvious relationship is visible with

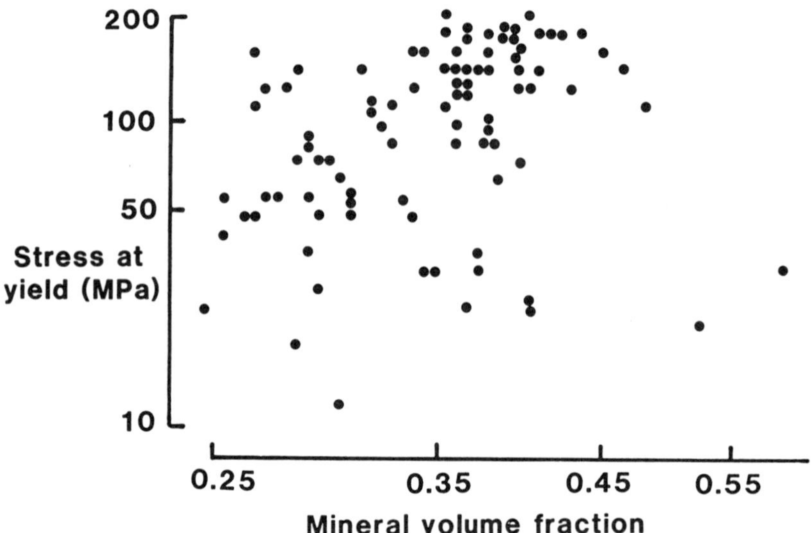

Figure 3. Stress at yield as a function of mineral volume fraction. (Mineral volume fraction is essentially calcium content expressed in volumetric terms.) Note log scale on ordinate.

mineralization, but apparent density appears to have a positive relationship.

Compared with its effect on post-yield stress and strain, the effects of variation in mineral volume fraction on yield stress and strain (see below) are comparatively small, though real. We can examine statistically the effect of changing the mineral volume fraction from the least to the greatest value found in the present data set, excluding the bullae, which are peculiarly weak, while keeping the other explanatory variables constant (at their mean value). Yield strain is halved, decreasing from 0.0094 to 0.0045. On the other hand yield stress is doubled, increasing from 77 to 154 MPa.

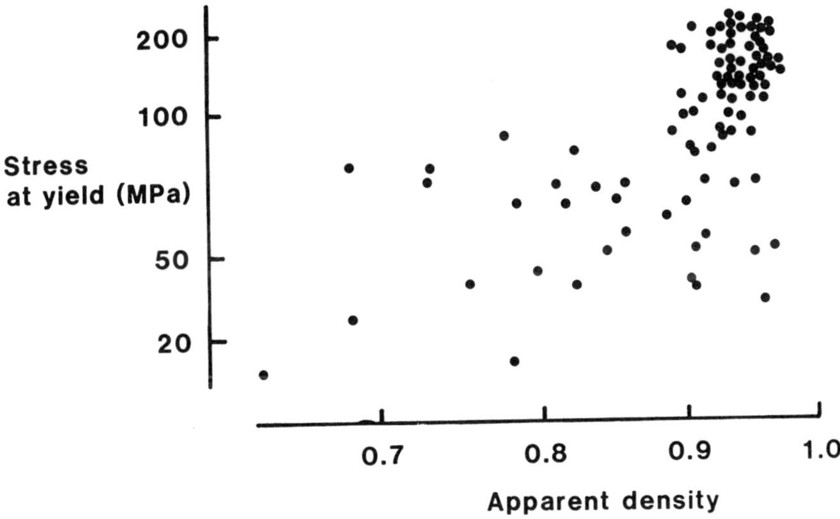

Figure 4. Stress at yield as a function of 'apparent density' (1- porosity). Note log scale on ordinate.

2.4 YIELD STRAIN AND ULTIMATE STRAIN

Figure 5 shows the relationship between mineral volume fraction and both yield strain and ultimate strain. There is a marked difference in the way these two mechanical properties vary in relation to mineral volume fraction. Yield strain shows a slight negative relationship, whereas ultimate strain shows a very clear and strong negative relationship.

A multiple linear regression using log mineral volume fraction, log apparent density, and orientation as explanatory variables, shows that only mineral volume fraction has a significant effect on ultimate strain. The value of R^2 shows that 74% of the variance is 'explained' by the regression. It is possible that the two extreme values (of the bulla), on the bottom right of Figure 5, might be having an undue influence on the form of the regression. However, if they are omitted from the data set the results are little changed.

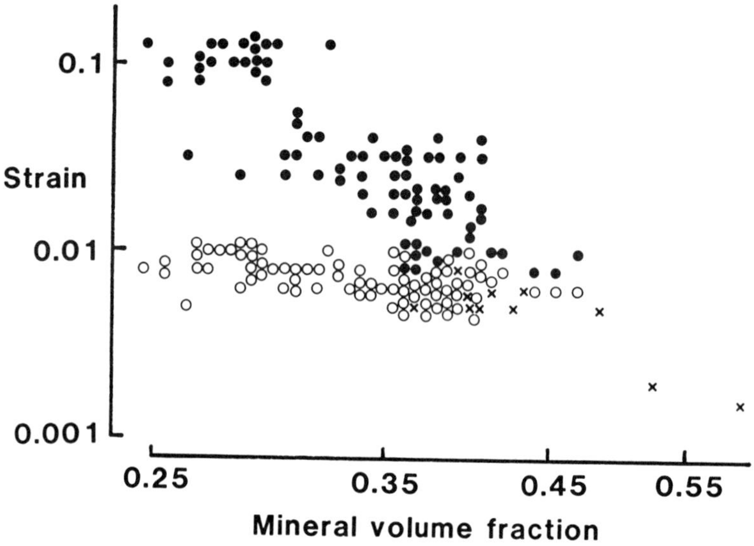

Figure 5. Yield strain and ultimate strain as a function of mineral volume fraction. Note log scales on ordinate and abscissa. Closed circles: ultimate strain; open circles: yield strain; crosses: specimens in which no post-yield strain was present.

The coefficient for mineral volume fraction is barely altered, from -5.16 to -5.08. R^2 is only slightly reduced, from 74% to 70%.

The large fracture strain associated with a low mineral volume fraction is produced by generalized strain occurring throughout the specimen, and not just by processes occurring near to the fracture. Unlike the situation in many metals, no necking occurs in bone specimens. The cross-sectional dimensions are virtually unaltered after fracture, even in those specimens that had shown 10% or so of ultimate strain.

The fracture surface, though often rough, particularly in the specimens that showed large strains at failure, shows no very long pullouts or tendrils, of the order of length of a millimeter, such as would appear if the large strains were associated with crack travel during fracture. The large ultimate strains associated with lower mineral volume fraction must, therefore, occur diffusely throughout the specimen. The largest strains are also associated with large increases with post-yield stress. A clear difference between yield strain and ultimate strain is that ultimate strain is far more variable. Yield strain varies from 0.0045 to 0.0108, a factor of x 2.4, while ultimate strain varies from 0.0045 to 0.1340, a factor of x 28.5. Statistically, this variation in ultimate strain is very highly negatively correlated with the mineral content of the specimen.

2.5 ULTIMATE STRESS

For most bone, ultimate tensile stress is only slightly greater than yield stress, and so one might expect the relationship between the explanatory variables and both ultimate and yield stress to be about the same. However, many less mineralized specimens, particularly of antler, showed a considerable post-yield increase in stress (as well as strain). As a result there is no clear relationship between mineral volume fraction and ultimate tensile strength, nor does bringing in other possible explanatory variables help statistically. The behavior of the least mineralized specimens and the highly mineralized specimens differs greatly in the post-yield region. In highly mineralized specimens the stress-strain curve is almost flat in the post-yield region. In the less mineralized bones the stress continues to increase quite markedly with the strain. This is carried to an extreme in antlers, which as a result can sometimes have a very high ultimate stress, although their yield stress is much lower than that of highly mineralized bone. Figure 6 shows characteristic stress-strain curves of antler and bovine femur.

This behavior of relatively poorly mineralized bone is difficult to reconcile with the models of, for instance, Carter and Caler,[4] corroborated by Currey,[5] and of Fondrk *et al.*[6] who all suppose that most of the strain occurring in the post-yield region represents increased compliance produced by the accumulation of damage. Such models, which are based on creep and fatigue experiments, assume that damage accumulates at a *rate* proportional to a high power of the stress (the eighteenth power in the case of Carter and Caler's model).[4] This type of model explains well the top of the stress-strain curve of well-mineralized bone because, once damage starts to accumulate at a measurable rate and shows as an increase in compliance, then a small increase in stress will produce a very large increase in the rate of damage accumulation, and consequently a large

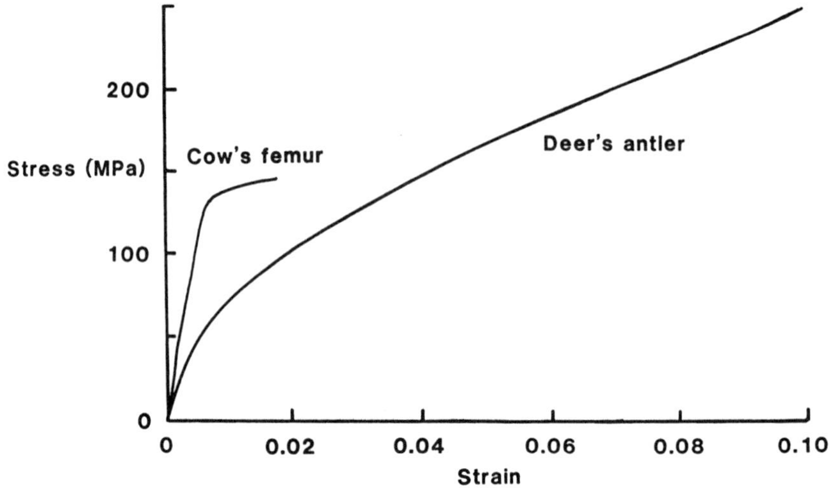

Figure 6. Typical stress-strain curves in tensile loading for bovine bone and red deer antler. Many antler specimens have the same general shape of curve as this, but fail at a much lower stress.

increase in the compliance. For instance, if the exponent relating stress to the rate of damage accumulation is eighteen, and an increase in compliance becomes noticeable at a stress of 100 MPa, then when the stress is 120 MPa the rate of damage accumulation will be 27 times greater; roughly a 1% increase in the current stress level produces a 20% increase in current damage rate.

The load-deformation curves of antlers initially have a fairly steeply rising portion, and a long nearly final straight portion, in which the compliance increases only slowly. But the region in between is gently curved (Figure 6). If the transition from the initial steep region to the curved portion represents increasing compliance caused by the accumulation of damage, then the kind of model proposed by Carter and Caler[4] cannot be valid for such bone.

The tympanic bulla of the whale, whose composition is probably typical of mammalian ear bones, has fracture properties that take it out of the mainstream of the relationships that are found in the other bones. Not only is it brittle, as might be expected of such highly mineralized bone, but its static strength is very low. The fracture surfaces of bones of various species named in Table-I show considerable variation in roughness. However, the surfaces of the bulla specimens are particularly smooth. There seem to be no mechanisms for arresting cracks; when a crack starts to travel in a bulla specimen it travels easily across the specimen. The change in mineral volume fraction, from 0.28 in the red deer antler, having a tensile strength of 250 MPa, to 0.56 in the whale's bulla, having a tensile strength of 27 MPa, has produced a *qualitative* change in the mode of fracture.

2.6 CREEP RUPTURE

As mentioned above, Carter and Caler[4] have produced evidence, based on the stress-rate dependence of failure stress, that the failure process in bone can be considered to be the result of the accumulation of little damage events whose rate of occurrence is some very high power of the imposed stress. Currey[5] produced some evidence corroborating this. Failure occurs when the bone accumulates a certain amount of damage. Of course, the bone actually fractures across at one level only, rather than through the whole specimen, and so one must suppose that the bone fractures when the general level of damage is so high that one particular cross-section, by chance having a greater amount of damage, or more initial voids than the surrounding bone, can no longer bear the load. In a later paper Caler and Carter[7] suggested that σ/E (stress divided by Young's modulus) was a more useful explanatory variable than stress. In the linearly elastic region σ/E is equivalent to strain, but in the post-yield region the relationship between stress and strain is complex. Another way of examining this phenomenon is to load specimens to a constant load, and measure the time to failure. The time to failure is assumed to be a negative function of rate of damage accumulation. Caler and Carter[7] performed such experiments on human bone, and obtained results consistent with their damage accumulation model. We performed such experiments on bovine bone, and also the antler of red deer, which is less highly mineralized (Figure 7).[8]

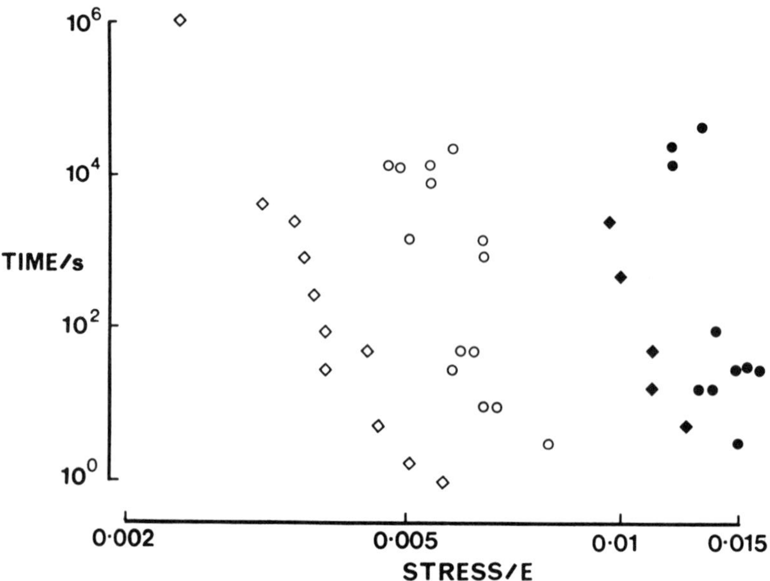

Figure 7. Creep rupture experiments. Time to failure, in seconds, as a function of Stress/Young's Modulus. Open lozenges: data from Carter and Caler, of human bone at 37°C; open circles: bovine tibia; closed lozenges: antler tip; closed circles: antler base. Note log scales.

The results show that there are great differences in the creep rupture properties of antler compared with 'ordinary' bone. All three materials examined showed a reduction in the time to failure as σ/E is increased. The relationships are not clean, and there is no evidence that the *slopes* of the lines are different, although their positions certainly are. Antler takes far longer to rupture at any given values of σ/E than does bone. However, if *stress* is taken as the explanatory variable, the evidence for a negative relationship between time to failure and stress is much weaker, and disappears in the case of bovine bone.

Figure 8 shows that, in bones taken as a whole, there is no relationship between maintained stress and time to fracture. There is, however, a clear indication that the more highly mineralized bone survives longer at any stress than the less mineralized antler

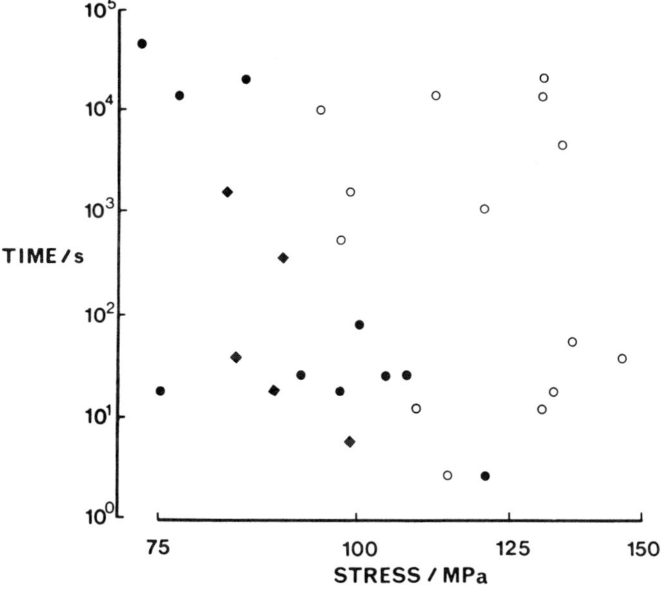

Figure 8. Creep rupture experiments. Time to failure as a function of stress. Values for Carter and Caler's experiments not included here (Ref. 7). Symbolization as in Figure 7. Note log scales.

bone. Since our data show that there is indeed a clear relationship between time to fracture and σ/E (which is effectively some measure of the initial strain) the problem is: why is there no relationship overall? The cumulative damage model of Caler and Carter[7] supposes that damage accumulates as some high power of stress or, assuming that the material is linear, of strain. Our experiments show first, that it is strain, not stress that is the important variable and, second, that bones of different degrees of mineralization show markedly different responses to strain: strain that would cause a highly mineralized specimen to accumulate damage at a very high rate leaves less mineralized bone unaffected. Although the cumulative damage model is in many ways a very satisfactory one, to be comprehensive it needs to take into account the markedly differing responses to strain of bones of different degrees of mineralization. Experiments on bone material with mineralization intermediate between those of antler and bovine bone should give insight into this phenomenon.

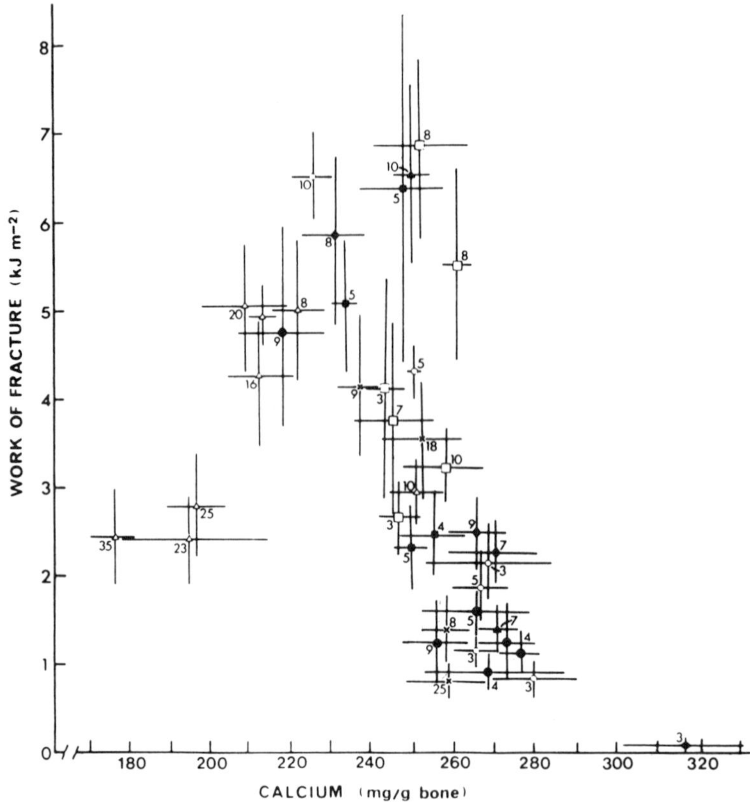

Figure 9. Species means of work of fracture as a function of calcium content. (Bars represent ± two standard errors.) Numbers in the body of the diagram represent mean porosity (note very large porosity of the three lefthand species). Symbols represent categories not discussed in this review.

2.7 TOUGHNESS

There is rather little information about the fracture mechanics properties of bones.[9] The sizes and shapes in which bones occur make it difficult to perform fracture mechanics tests. We do have a fairly large number of values of the work of fracture of different types of bone, in which the dimensions of the test pieces were always the same, thus obviating size effects. (Work of fracture is the work under the load-deformation curve of specimens broken statically in three-point bending, the specimens being specially shaped so that the crack always runs in a controlled way. We have divided this work by twice the cross-sectional area of the fracture surface.)[10]

Figure 9 shows mean values of work of fracture as a function of calcium content for *species.* Although the points describe sets of data each of which can be rather varied, the trend is nevertheless clear. At the lowest values of mineralization the work of fracture is rather low (though the picture is somewhat confused by the fact that these low calcium bones also have rather high porosities). It rises rapidly and then falls again. If the load-deformation curves are examined, it is found that the very poorly mineralized specimens reach only a rather low load before the crack starts to travel, but that a fair amount of energy is absorbed in driving the crack through the specimen. Highly mineralized specimens, on the other hand, show rather little extra energy absorption after the highest load is reached, but this load is itself quite high, so the total energy absorbed is not trivial. Finally the specimens with a mineralization rather less than the mean (which are mostly antler) have both a reasonably high greatest load, and also absorb much energy during the crack propagation process.

Another measure of the toughness of a material is given by the area under the tensile load deformation curve. This is not very satisfactory in a number of ways, not the least because it is impossible to account for the energy stored in the testing machine which is lost as vibration when the specimen ruptures. Nevertheless, if the specimens are all the same size and shape, as here, it does give some idea of the ability of a specimen as a whole to absorb energy before rupture, even though stronger specimens, bearing higher loads before fracture, will appear somewhat more brittle because the higher loads will cause more energy to be stored in the machine. Generally the area under the curve falls rather dramatically with the amount of mineral in the specimen. This fall is caused almost entirely by the reduction in the post-yield strain.

3.0 DAMAGE

3.1 A DAMAGE PARADOX

The creep rupture experiments and the phenomenal toughness of some bones, such as antler, suggest that the ability to tolerate microdamage, and the various forms by which microdamage accumulate in bony tissues are crucial factors determining the ultimate properties of these structures.

Recently we have damaged bone and antler specimens in 3-point bending and induced diffuse microcracking over their central region. The microcracking caused by this treatment can be seen by staining techniques and it has been reported previously.[11] The specimens were flexed to progressively higher loads over a small number of cycles (<10). An average drop of 15-25% for the secant modulus (indicating the increased compliance of the damaged tissue) was produced in the bone specimens and an average drop of 30-40% in the antler specimens. One of the bone specimens broke in the 3-point bending damaging regime, showing that we had approached a limit for the amount of damage that bone could sustain before catastrophic failure occurred. On the other hand antler specimens flexed easily to higher displacements, conforming to their characteristically different tensile behavior.[3] We define damage, ω as:

$$\omega = 1 - (E/E_0) \tag{1}$$

where E_0 is the initial modulus, and E the present secant modulus.[12] The specimens were then fractured in simple tension by single pull to failure. Undamaged bone and antler specimens acted as controls, their stiffnesses and strengths being used for comparison with the specimens that were damaged by cyclic loading.

An average damage value of somewhat less than 0.2 had been inflicted on the bone specimens by bending and somewhat less than 0.4 on the antler specimens. The reduction in stiffness was irrecoverable. If the damage incurred by the bending specimens were to be shown in the tension tests, it might be thought that the antler would have a reduced modulus, and that although the reduction in its modulus would be greater than for that of the bone specimens, nevertheless the bone specimens would also show a considerable reduction in modulus. The average traces of the single pull to failure tests of all bone and antler specimens, both undamaged and damaged, are shown in Figure 10. In the antler the modulus of elasticity (measured in the initial linear region of the stress/strain curve) had been reduced, by 12.5% on the average (Table-II), but in bone the reduction in the modulus of the initial linear elastic region was small and insignificant. Clearly, the results on modulus derived from the bending tests do not at once translate numerically into differences in the simple tension tests, probably because damage is not uniform but concentrated in a central region, which has a much larger effect on specimen compliance in bending than in tension.

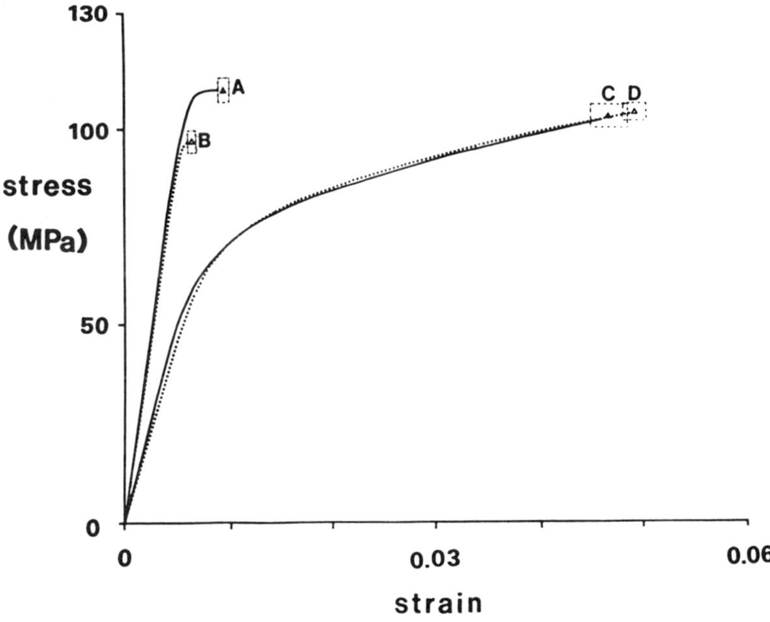

Figure 10. Stress strain curve for bone before (A) and after (B) being damaged in bending, and for antler before (C) and after (D) being damaged in bending. Curves are means, the boxes at each end represent ± one standard error.

In bone there was a significant reduction in strength and ultimate strain of the damaged specimens compared to the controls, but in antler, which had been damaged much more in bending, neither the strength nor the ultimate strain had been altered by the damage. This, at first sight, seems strange since it is widely believed that damage in general is responsible for a deterioration of a material's properties and its eventual failure. However, it is clear that more highly mineralized 'ordinary bone' is much more sensitive to the size, distribution, and density of cracks than is antler. The ability of the two materials to spread the damage is different. Although the antler showed a greater reduction in secant modulus, probably the damage producing this was spread more evenly along the specimen length, and so did not lead to problems when the specimen was loaded in tension. Antler shows a remarkable ability to perform satisfactorily in the presence of a significant amount of damage.[13]

Table-II. The relationship between damage in bending and the consequent effect on various mechanical properties, in bone and antler. The significance of the differences shown in parentheses.

Damage Estimate in Bending	Ultimate Tensile Strength	Ultimate Tensile Strain	Tensile Elastic Modulus
BONE:			
0	111.3 MPa	0.01	19.9 GPa
0.15 - 0.20	98.6 MPa (<0.01)	0.006 (<0.01)	19.3 GPa (NS)
ANTLER:			
0	102.9 MPa	0.047	11.2 GPa
0.35 - 0.40	103.5 MPa (NS)	0.050 (NS)	10.1 GPa (<0.05)

3.2 FRACTURE SURFACE ANALYZED BY FRACTALS

The fracture surface of vertebrate mineralized tissues gives some indication of their toughness. The fracture surface of tympanic bulla, which has an extremely low work of fracture is almost flat (Figure 11), whereas at low magnifications the fracture surface of antler, which has a very high work of fracture, is extremely convoluted, and shows many pull-outs and similar structures indicating a tearing mode of fracture.

We have quantified these differences using fractal analysis. Richardson plots were made of sections of the fracture surfaces of work of fracture specimens. Richardson plots involve measuring the length of the fracture line using a 'step length' of increasing size. As the step length increases, so the apparent distance between the two ends of the line will decrease, because the larger step length will miss all the little nooks and crannies which the smaller steps will discover. A plot of the log of the measured distance as a function of the log step length (the Richardson plot) will in general slope downwards as the step length increases. The more intricate the line the steeper will be the slope. If the surface is self-similar, that is, if the surface looks the same at all magnifications, the plot will be a straight line, of slope b between 0 and -1. The 'fractal dimension' D of the line will be given by $D = 1 - b$, and will have a value between 1 and 2. If the surface is not self-similar, that is if it seems smoother at some magnifications than at others, this will be reflected in the slope of the Richardson plot, and there will be no unique value of D, the fractal dimension.

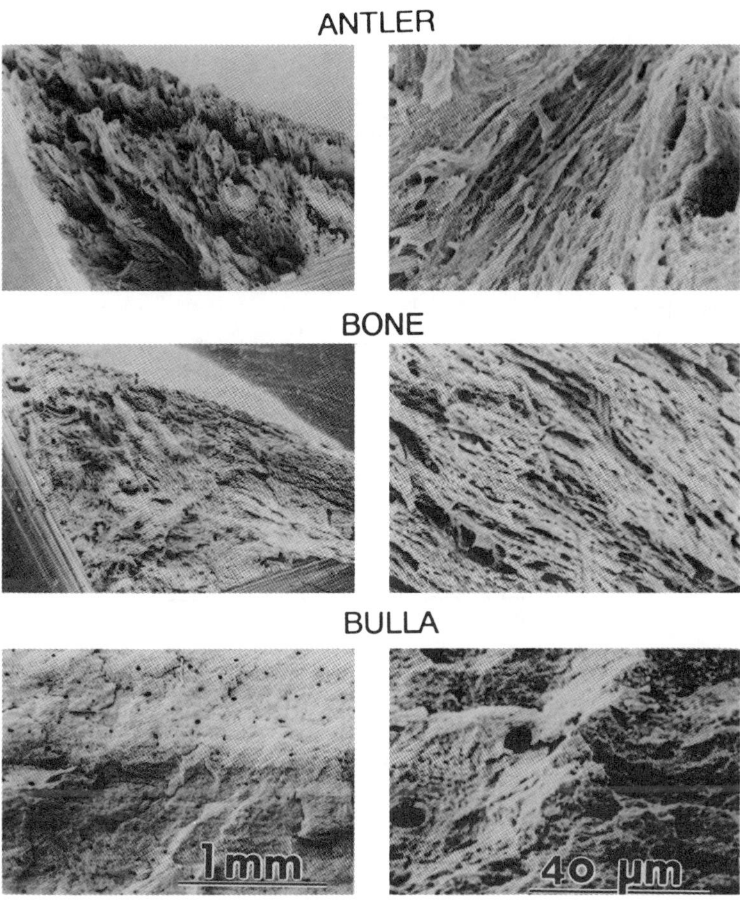

Figure 11. Fracture surfaces from work of fracture experiments at two magnifications. Red deer antler, work of fracture 3700 J m^{-2}.

Figure 12 shows the relationship between the local fractal dimension, varying as it does with the step length, and the work of fracture of specimens of bulla, antler, and 'ordinary bone.' In the fractal dimensions associated with lower magnifications (minimum step lengths greater than 40 μm) there is a positive relationship between work of fracture and the dimensions. The bulla is very smooth, with a fractal dimension of 1.05, while the very tough antlers and bone have values in the region of 1.25. As the step length decreases the situation alters. Except for bulla, which remains smooth at all step lengths, the low energy absorbing, highly mineralized specimens increase slightly

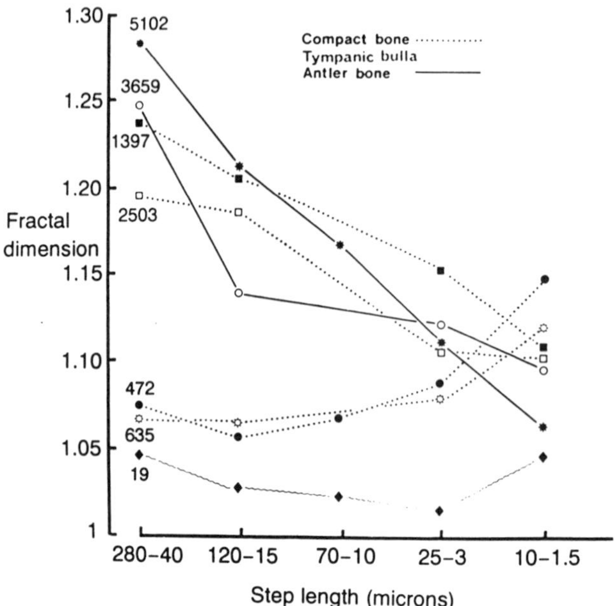

Figure 12. Relation between the fractal dimension and the work of fracture, at different step lengths. Ordinate: fractal dimension, calculated by regression, over various ranges of step length. Abscissa: range of step lengths used in calculating each regression. The numbers in the figure are the values for the work of fracture (joules per square meter) for the different specimens.

in fractal dimension, particularly at the shortest step lengths. However, the specimens with a work of fracture greater than 1000 J m^{-2} show a decrease in fractal dimension as the step length decreases. In the very highly mineralized bulla the fracture surface is very smooth at whatever scale it is viewed. In the two bones with low work of fracture the fracture surface is smooth at low magnifications, but becomes slightly rougher at a fine scale, below about 10 μm. On the other hand, the specimens with high works of fracture are very rough when seen at a low magnification, but are smoother when viewed at higher magnifications. Figure 13 shows the mean values of the measured lengths at different step lengths for two specimens, the wallaby with a work of fracture of 470 J m^{-2}, and the antler with a work of fracture of 3660 J m^{-2}. The way in which the slope of the curve varies with step length is obvious.

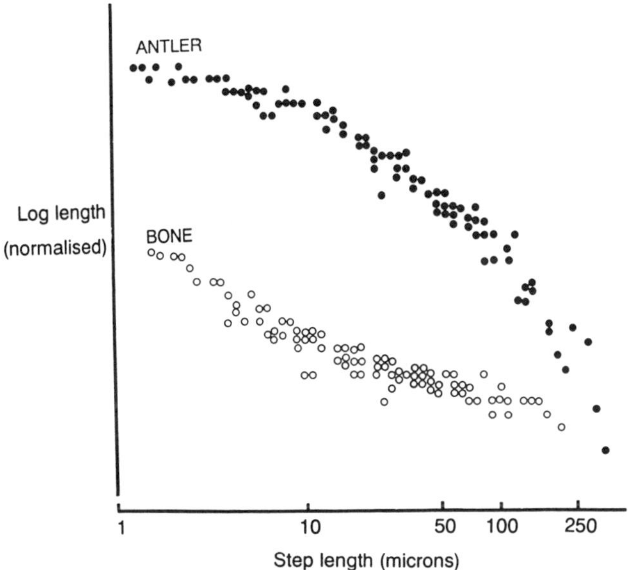

Figure 13. The mean values of the normalized measured length at different step lengths for an antler (filled circles) and a wallaby femur (open circles). The figure should not be taken completely at face value, because the distributions for measurements at different magnifications were shifted vertically to take account of the fact that the actual line being measured at different magnifications is necessarily different. The important feature is the general change in the slopes of the curves.

The fact that at high magnifications the fracture surface of the tough materials is rather smooth suggests that the components of these tissues peel away from each other rather easily (Figure 11, right hand column). (The 'components' are not collagen fibrils and mineral crystals, which are at a much lower level in the morphological hierarchy, but are bundles of fibers probably, often, bundles with different degrees of mineralization.) The optical results of Sedman et al.[14] discussed below show that, even in notched specimens, damage occurs quite diffusely, and the fatal crack does not develop until the optical changes that indicate microcracking have spread a considerable way across the specimen at the level at which the fatal crack will develop. These facts suggest that the peeling apart and separation occur deep within the specimen, before the fatal crack starts to run. Before the specimen breaks completely, the distances separating the new surfaces on each side of any crack will be very small, although large enough to scatter light, but cracks will be so pervasive as to require a considerable amount of energy for their production, and will produce a considerable increase in the compliance of the tissue. Nevertheless (and this is most important) the cracks seem to remain isolated from each other, and do not interact. That is, the presence of one microcrack does not, through the effects of stress concentration or whatever, induce the formation of other microcracks nearby. In antler this is shown by, amongst other things, the fact that it has the ability to bear a considerably increased load after damage starts to appear in it. There is a possibility that at high magnifications rather brittle bones show a slightly rougher surface than the tougher tissues, indicating that they have more cohesiveness (Figure 11).

We suggest that surfaces in poorly mineralized bone and antler pull apart easily, although requiring much work overall, and stress concentrations are blunted at the micro level. Furthermore, because the cracks form easily, but nevertheless require work for their formation, the strain energy stored in the specimen increases far less rapidly than the total work under the stress-strain curve. The strain energy required to drive a fatal crack is not available. Eventually the two parts of the bone simply peel apart from each other, absorbing a considerable amount of energy in this part of the fracture process also. On the other hand, the more highly mineralized bone forms cracks that are sharp, and though too small initially to develop into fatal cracks, are more potentially dangerous. Their spread is made easier by the cohesiveness of the bone, which will allow a greater accumulation of strain energy. As a result, although rather brittle bone has a respectable ability to accumulate micro-damage before rupturing, the microcracks start to interact at much lower strains than in the tougher tissues and a runaway confluence of microcracks results, not absorbing much energy in the actual fracture process.

The optical changes occurring around the tip of the macrocrack are indistinguishable from those occurring elsewhere in the specimen that do not progress to fatal cracks, suggesting that the fracture surface has morphological characteristics which are not in general unique to it, but are merely those of a set of microfractures that have joined up to form a fatal macrocrack. An implication of this is that in poorly mineralized bone and antler many of the cracks are oriented nearly parallel to the direction of loading.

The roughness of the surface of fractured mineralized tissues, as shown by their varying fractal dimensions, gives clear indications as to the processes occurring well away from the eventual fracture surface, and fractal analysis is a useful method of obtaining quantitative measures of the characteristics of fracture surfaces in mineralized tissues, and gives insights into the fracture process in these biological composite materials.

3.3 OPTICAL CHANGES AS A DAMAGE INDICATOR

There have been observations some years ago[11,15] of optical changes when bone "yields." This is almost certainly caused by microcracking, and is not analogous to the volume-increasing crazing seen in many polymers. We have been examining this in our laboratory by use of high speed video to capture these effects over a wide range of strain rates in tensile tests, using ordinary bone, antler, and narwhal tusk. The first appearance of the whitening (when viewed by incident light) occurs as the load-deformation curve enters the yield region, and irrecoverable damage is inflicted on the tissues. Although the damage is irrecoverable, the whitening is to a very large extent abolished when the load is reduced to zero. We find that antler in particular has the ability to undergo damage (as indicated by whitening) over a very large proportion of the gauge length (Figure 14). Antler must have some mechanism that allows microcracks to initiate, but not to spread.[14]

3.4 IMAGING OF CRACKS BY SCANNING CONFOCAL MICROSCOPY

We have used the scanning confocal microscope to visualize the three-dimensional shape of microcracks in bone and antler. Our results show that microcracking is indeed widespread in bone that has entered the yield region, and that the microcracks are indeed very small (Figure 15). Near stress concentrations, the cracking is quite spectacular (Figure 16) yet the specimen is nowhere near failure.

4.0 DISCUSSION

4.1 THE COHESIVENESS OF BONE AND ANTLER

What it is that determines the nature of the cohesiveness of the components of the tissue is not clear. It is surely related in some way to the differences in mineralization, but we are almost entirely ignorant of the nature of mineral-collagen bonding, or mineral-mineral bonding in these tissues. Ear bones, such as tympanic bulla and the auditory

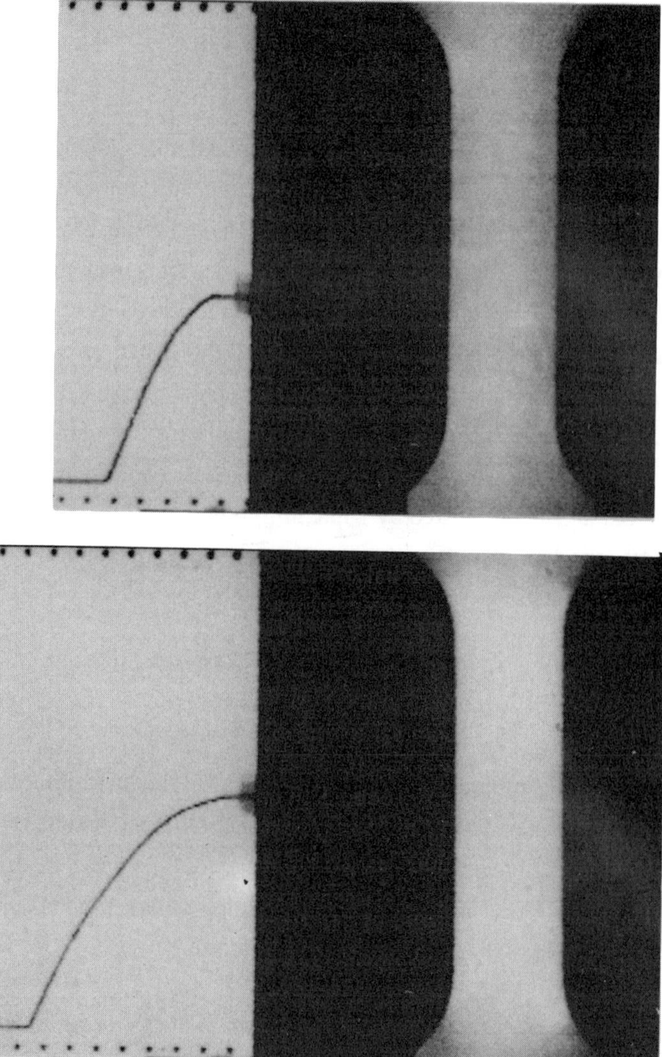

Figure 14. Whitening in antler loaded in tension. On the left is the pen recorder showing the trace of load and time. On the right is the specimen, seen by reflected light. Top, initial loading; the specimen has barely yielded, and is relatively dark. Bottom, the specimen has been loaded well into the yield region. It has gone white, except at the shoulder, where the strain is less.

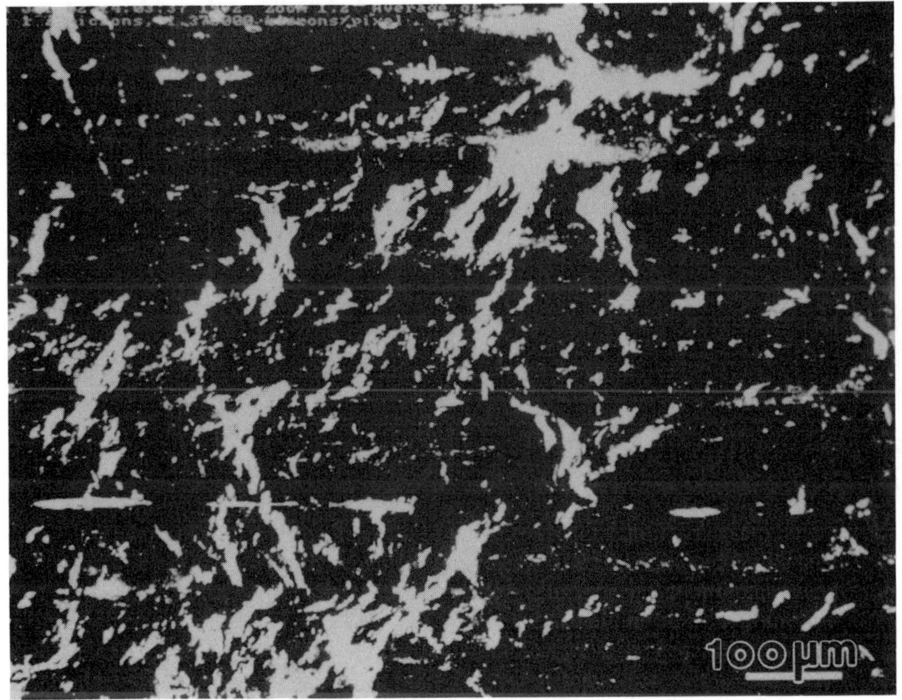

Figure 15. Scanning confocal microscope picture of bone that has been loaded just into the yield region. The microscope is imaging fluorescent dye that is excited by a laser beam. This is laminar bone, and the rows of bright lights (arrowed) are blood vessel cavities. The sloping line on the left is an artifact: a scratch. The other bright areas are cracks. This image is of the dye present through a depth of about 50 μm so the large white areas are not large voids, but cracks sloping with respect to the line of sight.

ossicles, are very heavily mineralized, are very brittle, and have rather smooth fracture surfaces, suggesting that highly mineralized tissues behave as a single block, and not as a composite. Somewhat less highly mineralized bones have some ability to develop non-fatal microcracks, but they cannot achieve high strains before the microcracks join up to form a fatal macrocrack. The least mineralized tissues, whether they are bone or antler, have the ability to form microcracks that do not spread or join up as strain increases. This ability greatly increases the toughness of the tissues.

Figure 16. As 15, but at a smaller scale, showing the distribution of cracks round a stress concentration. The laminar bone was loaded in tension. The rows of bright spots are blood channels. The diffuse stain on the right and left of the hole does not represent cracks, but is where dye has penetrated into the bone.

4.2 UNANSWERED QUESTIONS

This review has dealt mainly with toughening. It has to be said that, at the moment, the manner in which bone is able to be so tough is not clear. However, enough is now known for us to begin to get an idea of where to look. Many questions must be answered, but two are obviously of particular importance:

(i) What is the mechanism that allows all bone, except the most highly mineralized, to accept a considerable amount of microdamage, involving a considerable amount of energy dissipation, without the microdamage spreading right across the cross-section, and thereby causing failure?

(ii) What is the mechanism that allows lightly mineralized bone to undergo a considerable increase in stress after it has "yielded"?

At the moment, the ability of bony material to show these properties seems to depend to a large extent on its mineral content; roughly, the more the mineral the less the toughness. However, what we now need to understand are the structural changes, at the micro level, that changes in mineralization produce, and how they alter the micromechanics of bony toughening.

5.0 ACKNOWLEDGEMENTS

We are grateful to the U.S. Air Force Office of Scientific Research for supporting PZ with a grant to investigate toughening mechanisms in biological hard tissues, and the Science and Engineering Research Council for supporting AJS and for making available the high speed VCR and the Scanning Confocal Microscope. We are grateful to Professor John Tucker and his team at St. Andrews for letting us use the Scottish Confocal Microscope Facility, and helping in its operation.

6.0 REFERENCES

1. J. D. Currey, "The Effect of Porosity and Mineral Content on the Young's Modulus of Elasticity of Compact Bone," *J. Biomech.,* **21**, 131-139 (1988).

2. J. D. Currey, "The Evolution of the Mechanical Properties of Amniote Bone," *J. Biomech.,* **20**, 1035-1044 (1987).

3. J. D. Currey, "Physical Characteristics Affecting the Tensile Failure Properties of Compact Bone," *J. Biomech.*, **23**, 837-844 (1990).

4. D. R. Carter and W. E. Caler, "A Cumulative Damage Model for Bone Fracture," *J. Orthop. Res.*, **3**, 84-90 (1985).

5. J. D. Currey, "Strain rate and mineral content in fracture models of bone," *J. Orth. Res.*, **6**, 32-38 (1988).

6. M. Fondrk, E. Bahniuk, D.T. Davy, and C. Michaels, "Some Viscoplastic Characteristics of Bovine and Human Cortical Bone," *J. Biomech.,* **21**, 623-630 (1988).

7. W. E. Caler and D.R. Carter, "Bone Creep-Fatigue Damage Accumulation," *J. Biomech.,* **22,** 625-635 (1989).

8. M. Mauch, J. D. Currey, and A. J. Sedman, "Creep Fracture in Bones with Different Stiffnesses," *J. Biomech.*, **25**, 12-16 (1992).

9. J. C. Behiri and W. Bonfield, "Fracture Mechanics of Bone-the Effects of Density, Specimen Thickness and Crack Velocity on Longitudinal Fracture," *J. Biomech.,* **17**, 25-34 (1984).

10. H. G. Tattersall and G. Tappin, "The Work of Fracture and Its Measurement in Metals, Ceramics and Other Materials," *J. Mater. Sci.,* **1,** 296-301 (1966).

11. J. D. Currey and K. Brear, "Tensile Yield in Bone,"*Calcified Tiss. Res.,* **15**. 173-179 (1975).

12. G. M. Newaz and W. J. Walsh, "Interrelationship of Damage and Strain in Particulate Composites," *J. Composite Mater.,* **23**, 326-336 (1989).

13. P. Zioupos, A. J. Sedman, and J. D. Currey, "Development of Damage in Bone and Antler and its Evaluation," in *VIII Meeting of the European Society of Biomechanics*, 41 (European Society of Biomechanics, Rome, 1992).

14. A. J. Sedman, J. D. Currey, and P. Zioupos, "Optical Changes as an Indicator of Mechanical Damage in Bone and Antler," in *VIII Meeting of the European Society of Biomechanics*, 40 (European Society of Biomechanics, Rome, 1992).

15. A. H. Burstein, D. T. Reilly, and V. H. Frankel, "Failure Characteristics of Bone and Bone Tissue," in *Perspectives in Biomedical Engineering*, R.M. Kenedi, (ed.) (Macmillan, London, 1973) pp.131-134.

BIOMIMETIC CERAMICS AND HARD COMPOSITES

Paul Calvert

Department of Materials Science and Engineering,
Arizona Materials Laboratories, University of Arizona
Tucson, Arizona 85712, USA

Biological structural materials are all composites and have sophisticated microstructures which allow good properties to be realized by rather weak polymers and minerals. By coupling these microstructures with strong synthetic polymers and ceramics, we could hope to produce high performance materials. This article reviews the properties of hard biological materials to provide a basis for comparison with synthetics, and discusses how biomimetic composites and ceramics may be produced.

1.0 INTRODUCTION

Structural biological materials are all composites. The basic structural units include elastomers, strong polymers, and reinforcing minerals. These components are mixed in various proportions and with various architectures to produce a wide range of mechanical properties. Combinations of hardness with toughness and of softness with strength can give rise to mechanical responses which seem to be superior to the equivalent synthetic materials.

The superior sophistication is easily seen by a comparison between micrographs of the complex morphology of bone or shell and of the equivalent ceramics or filled polymers. It is less easy to prove a superiority of properties since the component materials of bone, collagen, and hydroxyapatite cannot be regarded as high performance materials when compared to carbon fiber and polyimide. We therefore must make comparisons on a basis which allows for the different starting properties of the component materials. Assuming that we can convince ourselves of a performance improvement which arises from the material architecture we should consider how to develop synthetic materials with equivalent microstructures.

A second case may be considered, that biological materials do not prove to be superior in strength, toughness or any simple combination of properties but are desirable structures because they integrate better with the overall mechanical design of the system. In this case we are suggesting that the ability to locally tune mechanical properties allows a stronger structure to be built than if the machine was constructed from simple materials joined at sharp interfaces. Biomimicry in this case would require a conjunction of materials science, processing, and design engineering that is unlikely to occur in the short term.

Our approach has been to accept that the superiority of biological structures will be demonstrated and to consider what methods can be used to reproduce them synthetically. We have concentrated on hard composites and ceramics produced by the precipitation of inorganic reinforcing particles into a polymer matrix. There clearly are other approaches, ranging from perversion of a truly biological system to produce more useful materials to the use of existing processing methods to form more complex structures. In this paper I will review the features which we would like to see in a true mimicry of biomineralization and the progress which we have made towards this.

2.0 STRUCTURES OF BIOLOGICAL AND SYNTHETIC MATERIALS

The structural biological polymers include the polysaccharides, cellulose and chitin and the proteins, silk, elastin, and collagen. Cellulose and silk are highly oriented fibers with properties that are comparable with highly drawn Nylon or polyester fibers. The biological fibers are made through aqueous chemistry and are consequently much more hydrophilic, and much more plasticized by water, than are the synthetics. This water sensitivity has of course been the source of most of the drawbacks that wood has as a construction material, particularly its tendency to dry and split. An equivalent problem for the synthetic fibers is that melt spun fibers are limited to a melting point at which degradation can be avoided and so inter-chain bonding is kept relatively weak, compared to the high levels of hydrogen bonding in cellulose or silks. Silk is used primarily as a fiber, but cellulose and collagen occur in bulk though always basically in fiber form. It is interesting that biology has not used the strong polymers in unoriented form, in which synthetic polymers occur in molded plastics, and that biology does use continuous polymer fiber reinforcements but not mineral fiber reinforcements.

Collagen is of particular interest because the chains are wound into triple helices of about 640 nm length which are hydrogen bonded into fibrils. This discontinuous structure reduces the fiber strength when compared to silk or cellulose.

If we now compare the matrix phase of bone with that of the epoxy in a carbon-fiber composite, we notice important structural differences. Dense bone is about 40 vol% mineral, 20 vol% water, and 40 vol% collagen. The collagen is essentially discrete, highly oriented fibers in a gel matrix. This is a soft but strong structure. In contrast, epoxy resin is usually highly cross-linked and relatively brittle (Figure 1). The toughness of the composite essentially derives solely from the weakness of the interfaces. Current trends in composites are to use thermoplastic matrices with better toughness but the constraints imposed by the need to infiltrate the polymer amongst the fibers drastically limits the molecular weight of the resin.

Thus if we set the water-sensitivity of biopolymers aside, there are clear parallels in structure and performance between the various biological polymers and the main groups of synthetics, but they nonetheless are often used in different ways.

If we turn our attention to the reinforcing fillers the differences are very clear. The main biological reinforcements are calcium carbonate and hydroxyapatite. These relatively soft (Mohr hardness 3 and 5) compounds contrast with structural synthetic materials with higher hardness and internal bond strength such as silica, alumina, silicon carbide, and graphite. Harder materials are available to organisms, magnetite and silica do appear, but the improvement is clearly not sufficient to justify silica bones or magnetite shells. We can view this as an example of incompetence in the evolutionary system, as a reflection of metabolic costs dominating structural design or as a result of some improved control of precipitation and resorption for the crystalline calcium salts when compared to the oxides.

The form of the reinforcing phase shows substantial differences between natural and synthetic composites. Typical continuous fiber reinforcements have diameters of 10 microns. Many powders are used for particulate reinforcement of polymers but typical sizes are in the range of 0.1 microns to 20 microns. Bone is reinforced by platelets with a thickness of 4 nm, a width of 30-50 nm, and a length of 100-500 nm. This axial ratio,

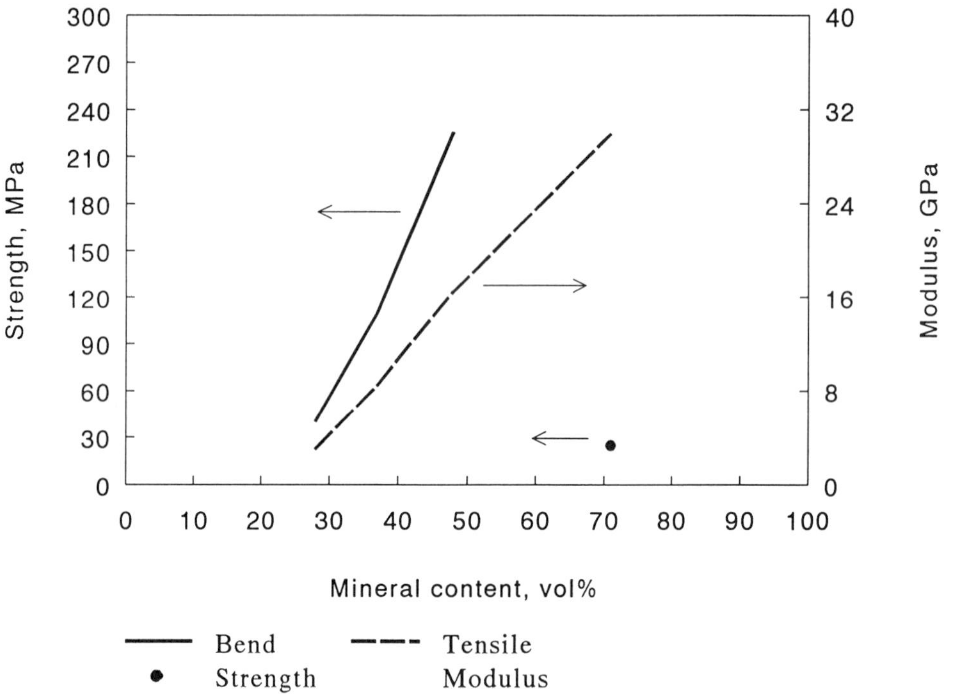

Figure 1. Properties of bone vs. mineral content.

of 20:1 or more, would be expected to give good load transfer from the matrix to the platelet if the interfacial bonding is strong.

Tooth and shell have such low organic contents that they are best thought of as ceramics. In enamel the hydroxyapatite crystals are about 40 nm wide by 150 nm long and bundled into larger rods, in shell the crystals have thicknesses of 0.5 μm by several microns wide. This scale is comparable with the sizes of the finest synthetic ceramics. The key differences lie in the presence of the organic layer surrounding the crystals and in the highly anisotropic shape of the crystals. Both of these factors may contribute to the higher toughness of natural ceramics when compared to synthetics. It is noteworthy that small dentine crystals and large enamel crystals can be formed simultaneously and within a short distance of one another, controlled by local protein concentrations.

3.0 SYNTHETIC AND BIOLOGICAL COMPOSITES

Bone is a composite material where a tough collagen matrix is reinforced with hydroxyapatite plates.[1] As such we would expect to be able compare the structure and mechanical properties of bone with those of equivalent reinforced polymer composites. Using theories for the modulus of composites with values for the moduli of the collagenous matrix and of hydroxyapatite, we can expect to predict the modulus of bone. This has been discussed by Currey in this book (page 115).

Table-I. Mechanical properties of bone and replacement materials for bone

	Tensile Modulus, GPa	**Bend Strength, MPa**	**K_c MPa $m^{-1/2}$**	**Toughness, G_c, Jm^{-2}**
Cortica 1 bone	7-30	50-150	2-12	600-5000
Alumina	365	6-55	3	40
Ti Alloy	106	900	80	10^4
PMMA	3.5	70	1.5	400
Polyethylene	1	30		8000
Polyethylene-Hydroxyapatite, 40 vol%*	5	22	3	1800
Polyethylene-Hydroxyapatite, 50 vol%	9	26	3	1000

*Bonfield et al.[13]

Strength is a less predictable property as it is more dependent on interfacial bonding and the detailed morphology of the composite. There have been efforts to reproduce the properties of bone for implant materials but, as shown in Table-I it is difficult to get a sufficiently high modulus without severely reducing the strength of analogous synthetic composites. This stems from the random packing which leads to fiber-fiber contacts and provides an easy fracture path.

Table-II. Natural and synthetic composites

	Elastic Modulus, GPa	Strength, MPa	Work of Fracture, kJ m^{-2}	Fracture toughness, K_{1c} $MPa \cdot m^{-1/2}$	Strain to break, %
Bone (bovine femur)	20	220	1.7	5	10
Insect cuticle (35 vol% chitin fibers)	6-10	80			
PEEK-Carbon fiber, 61 vol%, Longitudinal	140	2200	1.6		1
PEEK-Carbon fiber, 61 vol%, Transverse	8.3	73			1
Polyethyleneterephthalate + 18 vol% glass fiber	9	144			6.6
Polyethyleneterephthalate + 35 vol% glass fiber	20	165	3.2	9.5	1
Sheet molding compound (35% glass fiber)	16	158			1.7
E Glass	70	3,000			
Hydroxyapatite	130	100			
Polyethyleneterephthalate	3.3	60	7.3	4.9	275
Collagen (tendon)	2	100			

Data on synthetic composites from Reference 14.

Table-II compares bone with continuous fiber composites and with short fiber composites. Continuous fiber composites are very stiff and strong along the fiber axis but the properties are poor perpendicular to the fibers. In practice they also are extremely costly to process. Short fiber composites can be readily molded and are isotropic, but the properties are poorer. Bone seems to lie between these two groups. If we compare the structure of synthetic short fiber composites with bone, we see the following differences: bone mineral platelets are much finer than synthetic fibers, the packing densities are higher, interfacial bond strength is unknown but is probably similar to or higher than the matrix strength, the matrix is soft and tough rather than brittle, the filler is platy rather than fibrous and is highly aligned. The matrix is itself anisotropic and is oriented parallel to the filler. On the other hand the aspect ratio is in the range of 20-50 which is similar to composites.[2] The other challenging feature of bone is the hierarchy of structures coupled with large variations in orientation and density across the bone. Some of the structure on the millimeter-centimeter scale, such as the Haversian system in human long bones may be due to reparability or damage limitation rather than performance.

Glass bead filling of synthetic polymers results in an increase in modulus and strength at the cost of a severe reduction in elongation to break. Reinforcement with glass fibers results in a significant increase in modulus and strength. Bigg[3] has recently reviewed fiber-filled thermoplastic composites. Strength and modulus can be predicted on the basis of the rule of mixtures. Both depend on aspect ratio, degree of orientation, and interfacial bonding as well as on the volume fractions and properties of the fibers and polymer. Aspect ratio is limited by the breakage of fibers that occurs during processing, especially at high volume fractions. As a result it is not practicable to use fiber volume fractions of more than about 0.3, beyond which strength starts to drop and the viscosity becomes too high for good molding. As shown in Figure 1, based on data from Currey,[4] bone shows a steady increase in strength to 50 vol% mineral. Beyond this, at about 70% there is effectively one set of data from the whale (ear) tympanic bulla, which shows high modulus but low strength. Since the function of this material is sound transmission rather than mechanical, it is difficult to know if it is representative of bone at high mineral fractions.

Bone is comparable in filler content and aspect ratio to the glass fiber filled polyethyleneterephthalate. It can be seen that the biggest difference in the two materials is the much larger extension to break of bone. This may well arise from the laminar structure of bone which provides weak planes. In this sense bone can exploit the hierarchy derived from a very fine, well bonded, filler combined with weak planes every few microns. In fiber composites we require the filler-matrix interface to be weak enough to deflect the fracture, yet strong enough to transfer load. These two functions seem to be separated in bone.

Sheet molding compound (SMC) is a composite of unsaturated polyester reinforced with long glass fibers and clay. It has a high modulus but at the cost of being formable only by bending the sheet before curing. This material is replacing metal for many car body panels. It can be seen that a bone-like material would be substantially better.

Beyond these direct comparisons, one would like to be able to estimate the properties of a bone built from stronger materials than are available to the organism. Collagen is intrinsically weaker than other oriented polymers because the chains take the form of short, stiff rods. The comparison, in the table, of fracture of oriented collagen in tendon with yielding of isotropic polyethyleneterephthalate masks this difference. Similarly the tabulated fracture strength for hydroxyapatite is measured on bulk samples and so is much lower than it would be for the fine reinforcing plates. Hydroxyapatite is ionically bonded and glass or carbon are covalent and so should be intrinsically stronger when everything else is equal. One would expect much improved properties if a bone-like structure could be reproduced in a structure which was a dense mesh of oriented polyamide reinforced by very fine glass fibers with a high axial ratio.

4.0 SYNTHETIC AND BIOLOGICAL CERAMICS

We can compare natural and synthetic ceramics, as in Table-III. In doing this we regard nacre and enamel as ceramics, despite a small volume fraction of polymer in the structure. The natural materials are based on calcium salts with ionic bonding that is intrinsically much weaker than that of the oxide ceramics such as alumina. The interest here is in what strength and toughness can be achieved when viewed in some sort of relation to the intrinsic bond strength or to the modulus. Taken in this context the high work of fracture numbers must be regarded as very impressive. It is possible to ascribe this toughness to either the polymer fraction or to the high aspect ratio of the rods of hydroxyapatite in enamel and the plates of aragonite in nacre. In this context it is interesting to consider jade as a fibrous ceramic which is not polymer bonded but still retains much of the toughness.

Jackson et al. have discussed composite models for the mechanical properties of nacre.[5] Modulus is discussed in terms of shear-lag models and strength and toughness in terms of pull-out of the aragonite platelets in their polymeric envelope. He shows in Table-IV that the strength and toughness are much reduced if the shell is dried to the point of embrittling the polymer. We have similarly observed that a frozen clam shell is much more brittle than one at room temperature. Using the same model one can predict the strength of a similar polymer-bound ceramic structure formed with synthetic materials such as alumina and an engineering thermoplastic. Several hypothetical examples are

Table-III. Natural and synthetic ceramics

	Modulus, GPa	Strength, MPa	Gc, J m^{-2}
Alumina	350	100-1000	7
Fused silica	72		9
Nacre	64	130	600-1240
Enamel	45	76	13-200
Jade (Jadeite)	205		94-120

Table-IV. Properties of nacre

Pinctada nacre in fracture across the shell plane[15]

	Young's Modulus, GPa	Tensile Strength, MPa	Toughness, Gc, J m^{-2}
Wet	64	130	1240
Dry	73	167	464
Dessicated			264

Table-V. Estimated properties of nacre-like composites

Polymer	Platelet	Volume Fraction, %	Young's Modulus, GPa	Tensile Strength, MPa
Polyphenylene sulfide	Alumina	95	151	319
Polyphenylene sulfide	Alumina	80	83	269
Polyphenylene sulfide	Silica	95	43	319
Epoxy	Alumina	95	151	133

Calculated following the model for nacre in Jackson et al.[15]
Assuming a platelet thickness of 0.5 μm and aspect ratio of 8.

given in Table-V.

The properties of these materials are impressive but applications as ceramics would be limited to below 300°C by the temperature sensitivity of the polymer. In tooth enamel there is no evidence that the small residual polymer fraction plays any mechanical role.

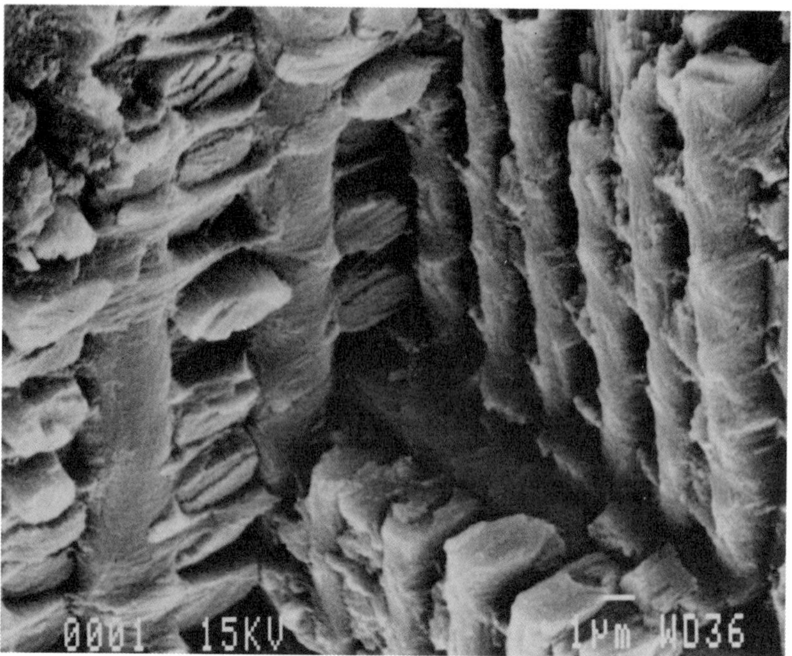

Figure 2. Rat tooth enamel showing crossed rods of hydroxyapatite.

The polymer does seem to be responsible for breaking up the hydroxyapatite rods (Figure 2) into bundles of fine fibers which presumably limits the effective maximum flaw size to well below 1 μm. The properties of enamel should thus be reproducible in a synthetic ceramic, if we can devise a way of reproducing this structure.

5.0 MECHANISMS OF BIOMINERALIZATION

Biological composites apparently form by a wide variety of different methods such that the only true principles governing their formation are rather general. All biological materials are made by aqueous processing at room temperature. The resultant structures are thus a product of direct chemical reaction rather than the result of thermally-induced changes.

If we focus on biomineralization, the processes clearly fall into two groups. Intracellular precipitation, such as sponge spicule formation, takes place within a vesicle surrounded by a lipid bilayer membrane. The magnetic particles in magnetotactic bacteria form similarly. In this case the membrane controls the influx of reagents. We know little enough about mechanisms of membrane transport and so cannot really guess how the conditions are established for precipitation within the vesicle. We would expect a combination of enzymes attached to the inside or outside membrane surface and transport proteins within the membrane.

The membrane may also act as a former to shape the growing particle, though sponge spicules often have a central polymer thread that is responsible for the initial shape formation.

Mineralization in synthetic lipid vesicles has been studied but the particles are usually grown by hydrogen ion, or other cation, transport into or out from the vesicle and there is little control over the deposition. Technically this process is unattractive because the low concentrations of vesicles and the low concentrations of ions in the vesicles severely limit the amount of material that can be formed. Scientifically we need more methods of control over ion transport and we need catalytic activity in the membrane before any interesting structures can be made.

The second group of mineralization processes are extracellular and so are more directly comparable to synthetic precipitation reactions. Tissues formed by this route include bone, tooth, and shell. In each case reagents must be delivered to a precipitation site and then the actual precipitation must be catalyzed locally.

Amongst the control mechanisms which may act in bone formation are nucleation sites, crystallization inhibitors, enzymes to break down inhibitors, enzymes to break down polyphosphates and create high concentrations of orthophosphate, impurity anions to control crystal size, and matrix orientation to control crystal orientation. Bone growth does also respond to local stresses but it is not known how the stress is sensed and transduced.

One clear ingredient in all these mineralizations is the presence of an extracellular matrix which structures the mineralization zone, prevents convection of reagents and particles, and allows localized concentrations.

Based on this analysis, we would like to be able to create composite and ceramic materials by precipitation within a polymeric matrix, and we would like to be able to demonstrate that such a matrix can specifically nucleate deposition, can have catalytic activity and can control orientation of the mineral. In addition we would like to be able to control particle size. With such a system we would hope to be able to form composites containing from 40-95 vol% of elongated sub-micron particles in a polymer. Such composites could be used as is, or sintered to a dense ceramic.

6.0 AVAILABLE SYNTHETIC ROUTES FOR IN-SITU CHEMISTRY

It is evident that mineralization of synthetic or biological polymers with calcium carbonate or hydroxyapatite, *in vitro* is quite feasible. Several groups are investigating precipitation of these compounds into proteins extracted from mollusks, into collagen and on variously treated synthetic surfaces. The approach can readily be extended to other sparingly soluble calcium carboxylates. We have prepared composites with potassium dihydrogenphosphate (KDP) in polyethyleneoxide by simple evaporation from an aqueous solution of polymer and salt. While such studies can teach us about the mechanism of biological mineralization, most of these materials are not intrinsically interesting. KDP is an exception because of its piezoelectric and non-linear optical properties.

Our own work has predominantly been concerned with the formation of oxide ceramic particles by hydrolysis of alkoxides from solution in polymers. We have grown particles of silica, titania, barium titanate, and zirconia. All these precipitates are initially amorphous but barium titanate, and presumably titania and zirconia, can be crystallized by treatment of the composite with boiling water.[6] This approach could obviously be extended to many other alkoxides. A number of metal chlorides such as iron (III) chloride and molybdenum (V) chloride, are highly soluble in organic solvents and can similarly be hydrolyzed *in situ* to oxides. In the case of iron, both magnetite and goethite can be formed.[7]

Reduction to metal *in situ* is also possible. Copper chloride has been reduced by sodium borohydride solutions to copper metal, as has iron(III) chloride. Sodium borohydride is not very penetrating and so tends to form surface films on polymers. Sodium naphthalide is more organophilic and will cause precipitation throughout a film. In the case of iron, hydrogen will reduce to the metal but the high temperature needed also tends to degrade the polymer. Sulfides are also readily attainable, either by direct reaction with hydrogen sulfide gas or by treatment of a metal-containing polymer with bis-(trimethylsilyl)sulfide.

It is reasonable to assume that this approach can be extended to anything that can be prepared by solution chemistry at less than 150°C in aqueous or organic solution, or from the vapor phase. So far, these constraints exclude many high temperature materials of interest to ceramists, such as the carbides, nitrides and borides. It also excludes silicon and most other important semiconductors. We have formed silicon carbide by an analogy of the rice hull process, whereby silica was precipitated into cellulose film that was subsequently carbonized under an inert atmosphere. Heating this carbon-silica composite to 1400°C produced a crop of silicon carbide whiskers on the carbon. The fact that the whiskers grow up, like grass, suggests that the deposition is via silicon monoxide vapor rather than by carbonization of silica. It is to be hoped that the range of materials formable *in situ* will continue to expand.

7.0 CONTROL MECHANISMS FOR PRECIPITATION IN POLYMERS

The characteristics of *in situ* precipitation in polymers that we would like to duplicate synthetically include high volume fraction of particles, elongated particles of normally equiaxed crystals, oriented particles, catalysis of precipitation by the polymer leading to localization of the reaction, site-specific nucleation in or on the polymer, and finally control of particle size.

Our studies of particle development during alkoxide hydrolysis in polymers show that phase separation can occur during drying of the polymer film, or during subsequent hydrolysis. If phase separation occurs of liquid alkoxide from an essentially dry polymer, the scale of the particles which form is about 0.5 μm. If the alkoxide is more compatible with the polymer, phase separation does not take place until hydrolysis occurs and the particle size is much reduced, leading to an essentially transparent film. This compatibility can be tuned by modifying the polarity of the alkoxide group. Hyd-

rolysis in moist air also can give smaller particles than drying followed by hydrolysis, because the large fraction of residual solvent compatibilizes the matrix.
We have investigated the catalysis of particle formation by two ethylene-(aminated acrylate) copolymers. The high concentration of tertiary amine groups in the amorphous regions of this polymer lead to a high internal pH in the polymer and a strong tendency to swell in acid solvents. When this polymer is placed in a solution of acidified methanol, tetraethoxysilane, and water, the polymer swells strongly and entraps silica. This response may be an essentially passive reaction to the solvent. If the same solution is first allowed to react for some days, it becomes viscous but shows no precipitation. On adding the solid polymer, there is an uptake of silicic acid by the polymer and precipitation starts both in the polymer and in the surrounding solution. This must reflect both movement of the silicic acid into the basic regions within the polymer and exchange of hydrogen ions between the polymer and the solution.

This system is catalytic in the sense that the polymer drives the precipitation reaction, but it probably does so through the medium of hydroxyl ions rather than through the influence of the chain itself. In past studies of gout, it has been found that sodium urate precipitation preferentially occurs within the highly polar matrix of cartilage rather than in other tissues, despite the activity of sodium urate being constant throughout much of the body.[8] Here again, the matrix seems to catalyze precipitation by providing a favorable environment. We need to give some careful thought to the classification of matrix-particle interactions to define what we really mean in thermodynamic and kinetic terms when we talk about matrix control of precipitation.

The strength and toughness of biological composites depend on reinforcement by elongated and oriented particles. To achieve such reinforcement by biomimetic *in situ* precipitation we must be able to grow rods or plates in a polymer matrix. We have shown that plates can be formed by drawing a polymer film containing partly precipitated hydrated titania.[9] The aspect ratio can be controlled up to about 20. The plate thickness is less than 1 μm. This elongation process could be incorporated into injection molding or extrusion of polymers. A true mimic of biological mineralization would allow elongated particles to be grown during treatment of a preformed polymer component. We have shown that a two phase polymer containing rods of polymethylmethacrylate in polyvinylidene fluoride can be mineralized with titania. The titania particles form as "sausages" or rows, like peas in a pod, in the polymethylmethacrylate zones.[10]

Biomineralization controls particle size, shape and orientation to form strong composites containing high volume fraction of mineral. We have demonstrated similar types of control for growth of inorganic oxides in synthetic polymers, but not yet simultaneously in a single sample.

8.0 APPLICATIONS OF BIOMIMETIC COMPOSITES AND CERAMICS

In the short term, one goal of biomimetic materials can be seen as devising ways of reproducing biological microstructures in order to develop enhanced properties in synthetic ceramics and composites. A second aim is to produce materials that will function as surgical implants with better biological compatibility.[11] In the longer term we can hope to produce complex sensing and actuating structures by the use of *in situ* mineralization methods. This can involve the precipitation of magnetic, piezoelectric, or electro-optical materials into polymers and will take us into the field of intelligent materials.

In composites, potential applications of this approach include moldable filled polymers with very fine scale reinforcement, polymer films with hardened surfaces, moldings where the reinforcement direction is elongated along directions of highest stress and transparent composites.

In ceramics, there are clear benefits from lamellar structures with fine, toughening polymer layers.[12] The use of polymer limits the maximum operating temperature to about 300°C, but it is evident that we can also expect considerable toughening from elongated ceramic grains particles or fibers without polymer. These structures may be prepared by conventional multilayer ceramic processing or by methods derived from biomineralization.

9.0 REFERENCES

1. *The Mechanical Adaptations of Bones*, J. D. Currey (Princeton University Press, Princeton, 1984).

2. *On Biomineralization*, H. A. Lowenstam and S. Weiner (Oxford University Press, Oxford, 1989).

3. D. M. Bigg, "Thermoplastic Matrix Composites," in *International Enclyclopedia of Composites*, S. M. Lee (ed.) (VCH Publishers, New York, 1991).

4. J. D. Currey, *Phil. Trans. R. Soc.* London, **B304**, 509-518 (1984).

5. A. P. Jackson, J. F. V. Vincent,146, and R. M. Turner, *Proc. R. Soc.* London, **B234**, 415 (1988).

6. P. Calvert and R. A. Broad, "Biomimetic Routes to Thin Ceramic Films," in *Materials Synthesis Utilizing Biological Processes*, P. C. Rieke, P. D. Calvert, and M. Alper (eds.) *Proc. MRS Symp.,* Vol. 174 (Materials Research Society, Pittsburgh, 1990) pp. 61-67.

7. P. Calvert and R. A. Broad, "Routes to Composites and Ceramics by *in situ* Precipitation in Polymers," in *Contemporary Topics in Polymer Science*, Vol. 6, W. M. Culbertson (ed.) (Plenum, New York, 1989).

8. *Crystals and Joint Disease*, P. A. Dieppe and P. D. Calvert (Chapman and Hall, New York, 1983).

9. J. W. Burdon and P. Calvert, "Growth of TiO_2 Particles within a Polymeric Matrix" in *Materials Synthesis Based on Biological Processes*, M. Alper, P. D. Calvert, R. B. Frankel, P. C. Rieke, and D. A. Tirrell (eds.) *Proc. MRS Symp.*, Vol. 218 (Materials Research Society, Pittsburgh, 1991) p. 203.

10. J. Burdon and P. Calvert, "Orientation in Biomimetic Polymer-Mineral Composites," in *Hierarchically Structured Materials*, I. A. Aksay, E. Baer, M. Sarikaya, and D. A. Tirrell (eds.) *Proc. MRS Symp.,* Vol. 255 (Materials Research Society, Pittsburgh, 1992) pp. 375-383.

11. A. H. Heuer, D. J. Fink, V. J. Laria, J. L. Arias, P. D. Calvert, K. Kendall, G. L. Messing, J. Blackwell, P. C. Rieke, D. H. Thompson, A. P. Wheeler, A. Veis, and A. I. Caplan, "Innovative Materials Processing Strategies: a Biomimetic Approach," *Science*, **255**, 1098 (1992).

12. M. Sarikaya and I. A. Aksay, in *Chemical Processing of Advanced Materials*, L. L. Hench and J. K. West (eds.) (Wiley, New York, 1992) p. 543.

13. W. Bonfield, M. D. Grynpas, A. E. Tully, J. Bowman, and J. Abram, *Biomaterials*, **2**, 185-191 (1981).

14. *International Encyclopedia of Composites*, S. M. Lee (ed.) (VCH Publishers, New York, 1991).

15. A. P. Jackson, J. F. V. Vincent, and R. M. Turner, *Proc. R. Soc.* London., **B234**, 415-440, (1988).

MICROSTRUCTURE OF AN INSECT CUTICLE AND APPLICATIONS TO ADVANCED COMPOSITES

Stephen L. Gunderson and Rebecca C. Schiavone

University of Dayton Research Institute
300 College Park, Dayton, Ohio 45469-0168, USA

Advanced composites require specific strength, specific stiffness, and damage tolerance as some of the characteristics in need of enhancement. Through the analysis of natural composite materials and structures and the application of their novel design characteristics to man-made systems (structural biomimetics), some or all of these challenges may be met. The bessbeetle, Odontotaenius disjunctus, cuticle is an excellent example of one such natural composite that possesses unique designs which may contribute to the development of improved man-made composite materials and structures. The bessbeetle cuticle is a natural structural polymeric composite that is lightweight, strong, stiff, and damage tolerant.[1] The microstructure, which is similar to that of its synthetic analog, consists of sheets (plies) of chitin fibers embedded in a proteinaceous matrix that are stacked on top of one another at various angles. Some of the novel designs in the bessbeetle cuticle which could be applicable to man-made composites include its structural hierarchy, fiber and matrix characteristics, constituent organization, and the methods by which it minimizes stress concentrations.

1.0 INTRODUCTION

A composite is a multiphase material of two or more distinctly different constituents which, when combined properly, provide improved performance over the individual components acting alone. Advanced fiber-reinforced polymeric composites (FRPC) consist of high-strength/high-stiffness, continuous organic or inorganic fibers embedded in a more ductile polymeric matrix. In a typical fabrication procedure, unidirectional fiber sheets (prepreg or tape) are stacked together with the fibers at various predetermined angles and processed under heat and pressure to form a laminate (Figure 1).

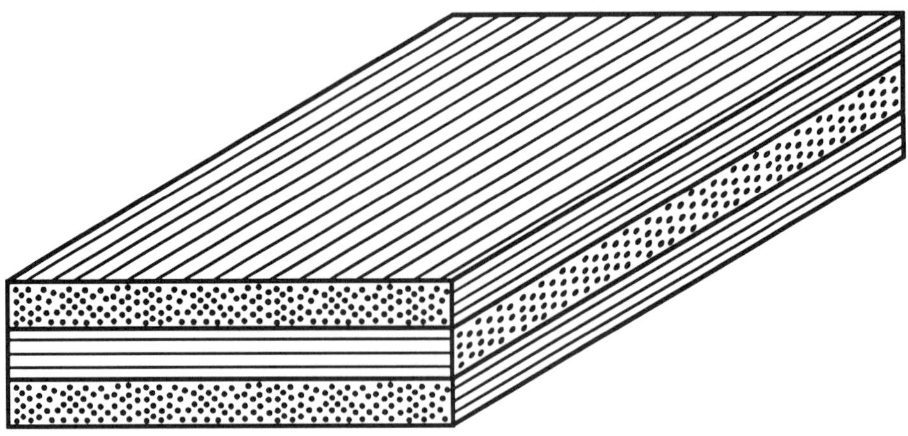

Figure 1. A schematic of three undirectional sheets or plies (two 0° and one 90°) compressed together to form a laminated FRPC.

The fibers, which are the primary load-bearing components in FRPC, are much stronger and stiffer in the axial direction than in the transverse direction. Therefore, the orientation of the fibers is critical to the types and magnitudes of loads the structure is capable of handling. Through computer modeling and other design techniques, fiber-reinforced composites can be tailored via constituent selection and organization, to meet desired performance characteristics. The matrix, in addition to holding the fibers together, protects and transmits loads to the fibers. Some FRPC matrices undergo polymerization (or curing) during processing, making them stronger, stiffer, and more resistant to moisture, but also more brittle.

It is because of their light weight, versatility, and tailorability that fiber-reinforced composites have seen increased use in applications ranging from tennis rackets to military aircraft. But there are some drawbacks that keep composites from being the material of choice for many more structural applications; the raw materials are expensive, and the damage tolerance and specific properties (property-to-density ratio) need improvement. By utilizing unique designs, less material could be used to provide adequate properties, reducing both part and processing costs. It is primarily in the areas of material use and interaction, design, and processing that the analysis of natural composite materials and structures can provide guidance for the production of synthetic composites with improved damage tolerance and specific properties.

The constituents of natural composites are relatively weak and simple, but nature is able to combine, organize, and treat the components in such a way as to create extremely complex and intricate systems that possess excellent performance characteristics. While man is limited more by design and processing and less by constituent material selection, nature is limited more by material selection and thus compensates for this deficiency by excelling at the design and processing levels. Natural materials and structures grow under ambient conditions with their formation finely controlled from the molecular building blocks to the final macrostructure, creating a structural hierarchy with precisely-controlled interactions between levels. In fact, it is very difficult to determine where one level ends and another begins, for the purpose of distinguishing between material, structure, and system. Some natural composites are subjected to external loads during formation allowing the structure to be designed/developed under stress and thus improving its *in-situ* characteristics.

The insect exoskeleton, or cuticle, is an excellent example of a natural, structural, fiber-reinforced, polymeric composite and consists of two primary sections: the epicuticle and the procuticle (Figure 2). The epicuticle is 0.1 to 10 μm thick and is composed primarily of waxes, lipids, and proteins. Its primary function is to act as a protective barrier against environmental conditions such as wind, water, and abrasion, similar to the protective coatings used on man-made advanced composite structures. Procuticle is a larger section, ranging in thickness from about 10 to 100 μm. It consists primarily of chitin fibers embedded in a proteinaceous matrix and provides structural shape and mechanical stability.[2] This section of the cuticle is the focus of the current paper.

The procuticle is subdivided into the exocuticle and the endocuticle. The matrix of the outer exocuticle contains sclerotized protein, i.e., protein which has undergone a cross-linking process similar to curing in man-made composites. Sclerotization makes the protein matrix stronger, stiffer, darker in color, and more insoluble in water and other solvents. Sclerotization also increases the matrix shear stiffness which allows for improved transmission of loads between fibers[3]. The method by which sclerotization occurs has been a subject of dispute for nearly 50 years, with one belief being that it is caused by the introduction of quinones or other chemicals, and the other that it is a

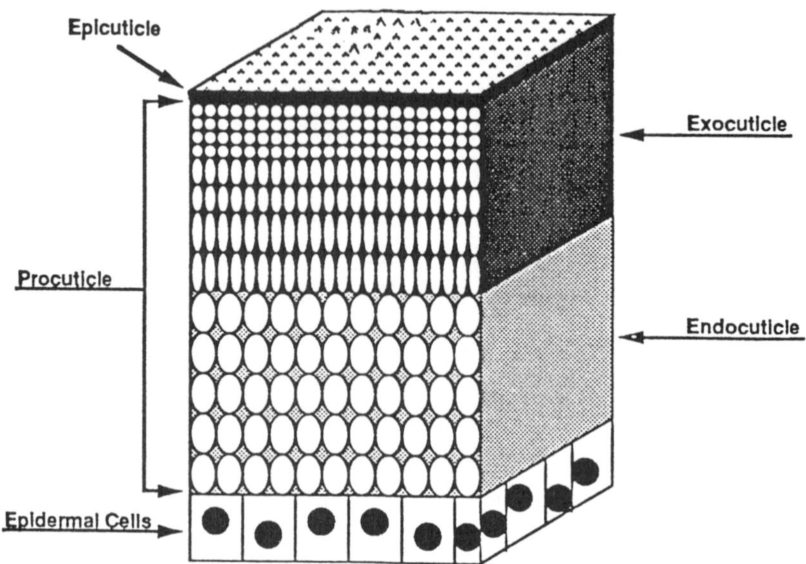

Figure 2. Schematic diagram of the insect cuticle displays its various divisions (not drawn to scale).

result of water loss. It could possibly be a combination of both mechanisms. The inner endocuticle contains an unsclerotized or slightly cross-linked matrix, making it softer and more ductile. The procuticle may contain completely sclerotized, completely unsclerotized matrix, or a combination of both; the last of which is the most common.

A related area of materials research is smart technologies which includes the areas of smart processing, smart structures, and smart skins. Smart technologies involve the use of sensors in a material or structure to retrieve processing information, monitor the health of the structure, and monitor external environmental conditions, all of which contribute to improved overall performance. Animals are innately composed of smart materials and structures that collect and process information on their well-being or health in real-time. For example, pain, thirst, hunger, body temperature, position of body parts, rate and direction of movement, blood pressure, pH balance, and levels of oxygen and carbon dioxide are just some of the information that are constantly monitored. At the same time an animal also processes information about its environment through visual, auditory, tactile, olfactory, and gustatory senses. The diversity of nature has produced many unique systems which help an organism survive under special conditions or in a specific niche. Studying these systems could assist in smart technology research and the production of man-made smart materials and structures.

2.0 STRUCTURAL ANALYSIS OF THE BESSBEETLE

The primary objective of this study is to improve the specific strength, specific modulus, and damage tolerance, and reduce stress concentrations in man-made advanced composites through the application of novel design concepts discovered from the analysis of the insect cuticle. The insect cuticle, specifically that of the bessbeetle (*Odontotaenius disjunctus*), was selected because it met the basic criteria of being a light-weight natural structural composite with good mechanical properties, was easily obtainable, and has a microstructure very similar to that of synthetic composites.[4]

The bessbeetle cuticle, or exoskeleton, is used primarily for protection against environmental conditions, predators, and wind-blown debris. Since the bessbeetle is a wood dweller, the cuticle must also be flexible enough to deform without breaking when burrowing. All stiff, fiber-reinforced natural composites, of which the bessbeetle cuticle is an example, are subjected to compressive, bending (tension-compression), and shearing loads.[3] This would explain the necessity for, and existence of, the cuticle's good strength, stiffness, and resistance to abrasion.

Sections of the bessbeetle exoskeleton from the leg, elytra, and pronotum were selected for study because of the possibility of discovering different microstructures (Figure 3). Aspects of interest for each section included composite cuticle thickness, number of plies, fiber orientation (layup), fiber and matrix characteristics, constituent interaction, and unique structural features. Methods of analysis included scanning electron microscopy (SEM) and optical microscopy (OM) in conjunction with various staining techniques. Finite-element models were also used to help explain and relate structure to function. It is important to note that the microstructure of insects varies between species and even within the same species, depending on growth conditions and habitat.[5] Therefore, the specific information presented here is not the same for all insect cuticles.

Figure 3. The bessbeetle, *Odontotaenius disjunctus,* with the sections of interest labeled.

2.1 STRUCTURAL HIERARCHY

Initial analysis of transverse cross-sections of the leg, elytra, and pronotum using SEM revealed similar microstructures: single rows of unidirectional chitin fibers embedded in a protein matrix forming thin sheets (plies) which were stacked on top of one another at various angles producing a laminated structure (Figure 4). Comparative cross-sections of man-made FRPC demonstrate microstructures that were very similar (Figure 5). One noticeable difference is the ratio of ply thickness to fiber diameter. A single ply in man-made FRPC contains between 15-20 fibers through the thickness, while in the bessbeetle cuticle each ply consists of a single row of fibers tightly packed together, a good example of the structural control Mother Nature possesses. Through matrix digestion with hot KOH, the amount of chitin present in the bessbeetle cuticle was estimated at about 40% by weight and 50% by volume. In comparison, man-made FRPC is about 70% by weight and 60% by volume.[6]

The protein matrix in the bessbeetle is initially soft and ductile but undergoes sclerotization making it stronger, stiffer, darker in color, and more brittle and hydrophobic. The interesting aspect of this process is that the bessbeetle has the

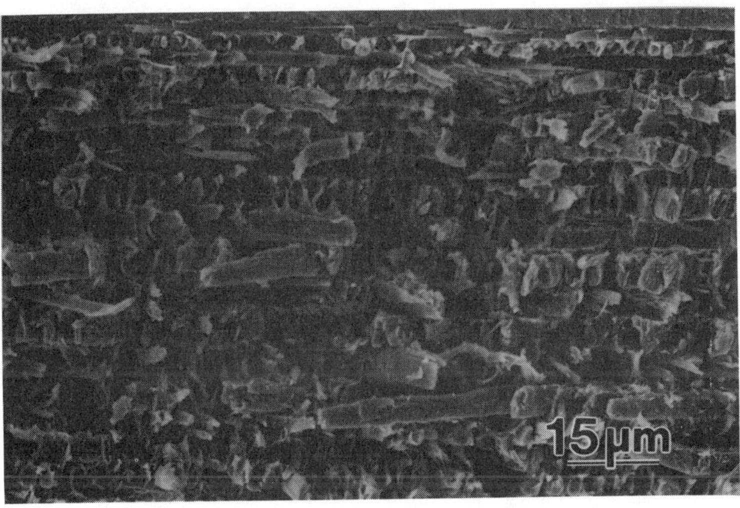

Figure 4. A transverse cross section of the *Odontotaenius disjunctus* cuticle. Note the laminated structure, variation in fibers, and precisely controlled microstructure.

Figure 5. (a) SEM image Close-up of the *Odontotaenius disjunctus* cuticle. (b) SEM image of a human made FRPC.

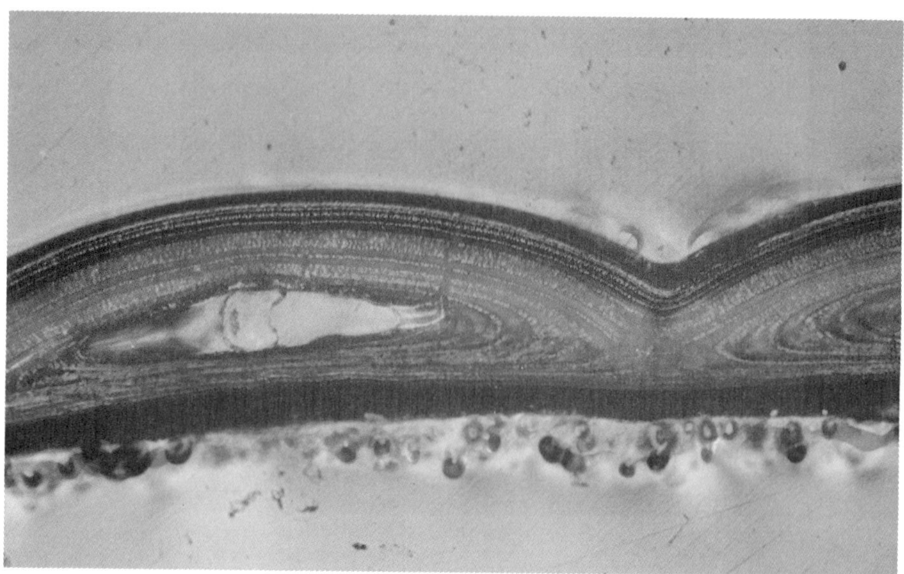

Figure 6. OM micrograph of a stained section of *Odontotaenius disjunctus* elytra cuticle exhibits the observed staining gradient. *(See also* ***Color Plate 3.****)*

ability, as do all insects, to vary the degree of sclerotization through the thickness of the cuticle. Through a combination of optical microscopy and staining techniques, a color gradient was observed through the thickness of the cuticle (Figure 6). This color gradient represents the relative cross-linking densities at various locations through the thickness of the cuticle with the outer regions being the most cross-linked. Theoretical studies indicate that the best design for a composite with a gradation in matrix crosslink density is to have the more crosslinked layer on the outside and the less crosslinked layer on the inside,[7] as is the case with most if not all insect cuticles. This type of arrangement improves the toughness, impact energy absorption, and bending stress and strain characteristics over the weighted average of the individual systems. Current research in the processing of FRPC includes attempts to monitor and control the degree of cure (cross-linking) of the matrix. If the technology is successful, it may then be possible to fabricate man-made composites with a gradient in matrix crosslink density similar to that observed in insect cuticles.

The structure of the elytra varies from those of the other sections studied;

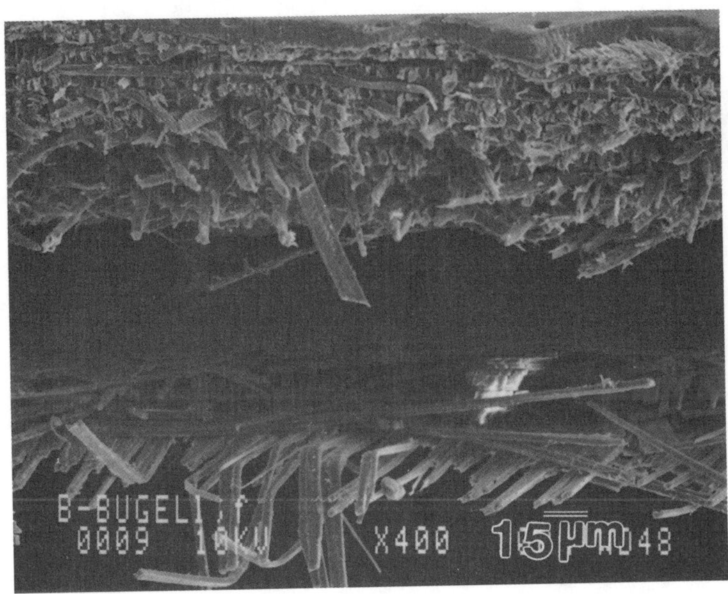

Figure 7. Transverse cross section of the *Odontotaenius disjunctus* elytra. Note the sandwich-like structure and the microstructural variations between the two halves.

it consists of two halves joined by a series of connectors, called trabeculae, producing a sandwich-like structure with a hollow core (Figure 7). The top half of the elytra is similar to the leg microstructure, but the bottom half appears to be made up of a series of laminated sheets or laths consisting of smaller chitin fibrils embedded in a protein matrix. The fibrils are unidirectional within each ply and the plies are oriented at various angles. The bottom half of the elytra more closely resembles man-made composites in that each ply consists of a number of fibers through the thickness, but on a much smaller scale (Figures 8 and 9).

The design of the elytra is similar to sandwich structures used by man in which two composite skins are bonded to a lightweight core. Sandwich structures provide good strength and stiffness coupled with excellent impact damage tolerance, while utilizing minimal composite material. This may explain the similar structural arrangement in the elytra. The elytra is the largest cuticle section and likely experiences the most "abuse." Therefore, the economical (with respect to the organism's metabolic cost) sandwich structure is used to provide strength, stiffness, flexibility, and impact damage tolerance using minimal material. From an examination of the microstructure it appears that the

Figure 8. High magnification micrograph of the bottom half of the *Odontotaenius disjunctus* elytra displays the well-known "plywood" architecture.

Figure 9. Low magnification micrograph of a synthetic laminated FRPC displays an appearance similar to that of the bottom half of the bessbeetle elytra.

bottom half of the elytra is the tougher, more damage tolerant of the two halves, acting as a final barrier to prevent cracks from reaching the inner delicate wings and epidermal sections.

The fiber orientation, or layup, in the bessbeetle cuticle is quite unique, and very similar in each of the three sections. This layup, which is termed a dual helicoid,[4] consists essentially of two orthogonal (0° and 90°) plies rotated through a (somewhat) constant angle through the thickness of the cuticle (Figures 10 and 11). This type of orientation was first discovered by Zelazny and Neville in 1972 while performing analyses on a number of large, heavily sclerotized beetle exoskeletons.[8] Man-made composite layups, as a general rule, are symmetrical and balanced to prevent warpage and the buildup of internal stresses. The dual helicoid orientation observed in the bessbeetle cuticle is neither balanced nor symmetrical. One possible explanation could be an attempt by the insect to achieve mechanical isotropy, i.e., the mechanical properties are independent of load direction.[9] Engineers, on the other hand, design composite structures based on predicted loading conditions.

Closer observations of the individual constituents revealed some interesting design concepts as well. The chitin fibers in the bessbeetle cuticle vary in size and cross-sectional geometry from ply to ply through the thickness[4] (refer to Figure 4). The major cross-sectional dimension of the fibers ranges from about 2-7 μm, while the geometry changes from circular to oblong. The fibers are small and circular in the outermost plies, larger and elliptical in the middle plies, then revert to circular geometry (without change in size) in the innermost plies. The elliptical fibers are consistently oriented with the major axis perpendicular to the cuticle surface. A computer modeling study revealed that specimens with such vertically oriented elliptical fibers are stiffer in bending and buckle at lower loads in transverse compression than specimens with horizontally oriented elliptical fibers. In addition, the use of small circular fibers on the compressive side of a specimen subjected to a bending load provides the best combination of stiffness and resistance to microcracking at the fiber/matrix interface. In light of these theoretical analyses, the bessbeetle cuticle (with its unique arrangement of fibers of different size and shape) has the flexibility to withstand lateral compression (as would occur when burrowing) and maximum strength and stiffness when compressed perpendicular to the cuticle surface. Therefore, the bessbeetle's combination of

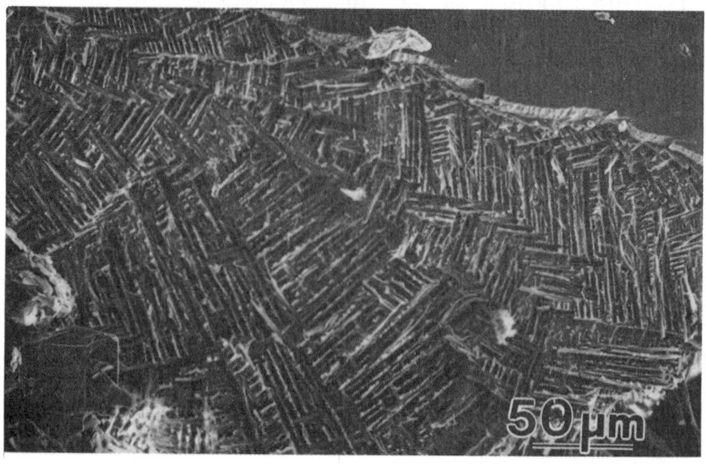

Figure 10. SEM micrograph of the fiber (ply) orientation observed in the *Odontotaenius disjunctus* cuticle.

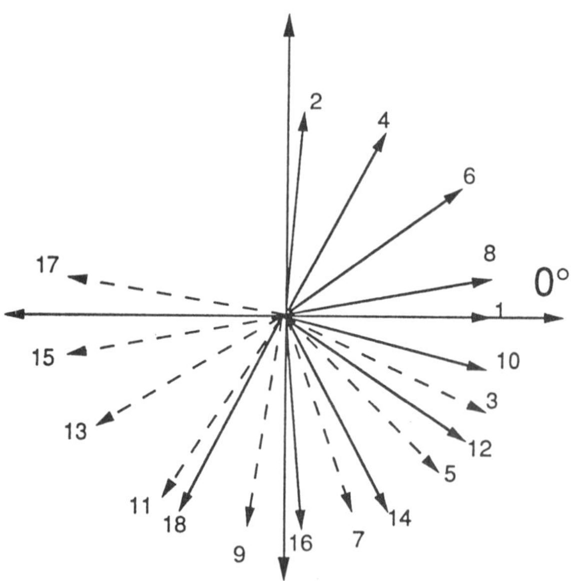

Figure 11. Schematic diagram of the dual helicoid ply orientation discovered in the *Odontotaenius disjunctus* cuticle.

different fiber sizes and shapes and the gradient in these parameters through the cuticle thickness may be an attempt to optimize each ply with respect to the magnitude and direction of loads incurred by the cuticle. Man-made fibers are generally of uniform size and shape throughout a composite, except in the case of hybrid composites where two or more *different* fibers are used. To our knowledge, no man-made composite has been fabricated from a single type of fiber where the fiber size and shape have been systematically varied.

2.2 MORPHOLOGY OF CONSTITUENTS

Each chitin fiber is composed of a bundle of fibrils held together by protein matrix (Figure 12). This fibrillar morphology closely resembles that of some man-made polymeric fibers such as Kevlar® and PBO (Figure 13). The chitin fibrils, which are approximately 3 nm in diameter, are composed of chains of N-acetylglucoseamine and are held together by covalent and hydrogen bonds. Some of the fibrils making up the larger fibers partially peel away, creating "furry fibers" which improves the interfacial bonding between fiber and matrix (Figure 14).

The protein matrix of the bessbeetle cuticle also has a fibrous morphology, as observed by high-resolution SEM (Figure 15), and resembles sheet molding compound (SMC), a man-made injection molded composite used in a wide range of applications. Whether these microfibrils are protein, chitin, or perhaps even some new material, is uncertain. A consequence of the fibrous nature of the matrix is improved damage tolerance which would help compensate for the brittleness brought on by the high crosslink density.

2.3 ADDITIONAL DESIGN CONSIDERATIONS

The fracture surfaces of failed cuticle sections reveal little or no fiber pull-out, indicating a strong interfacial bond between fiber and matrix. It is believed that beta-sheets in the protein interact with and bond to the chitin to form a stable and strong interface.[10] In addition some of the chitin fibrils even connect to adjacent fibers creating a network of fibril bridging[11] (Figure 16). This type of design improves a number of mechanical properties, primarily shear strength and damage tolerance. It is through the use of such toughening mechanisms as a fibrous matrix, "furry fibers," and fibril bridging that the

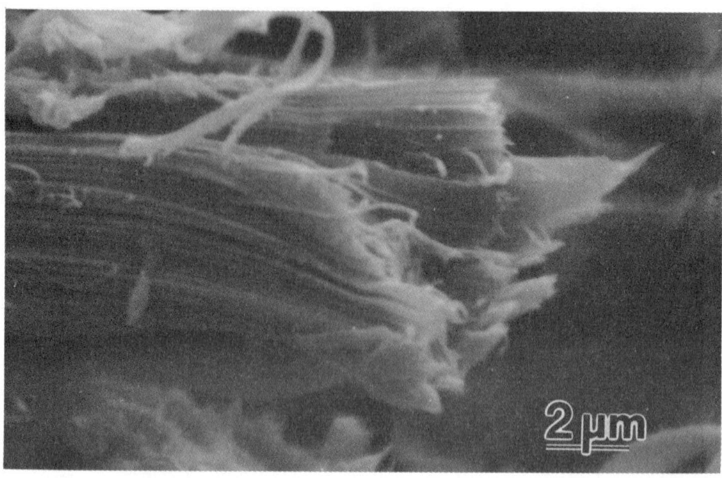

Figure 12. SEM micrograph of a failed chitin fiber displays the many microfibrils that are grouped together to form the fiber.

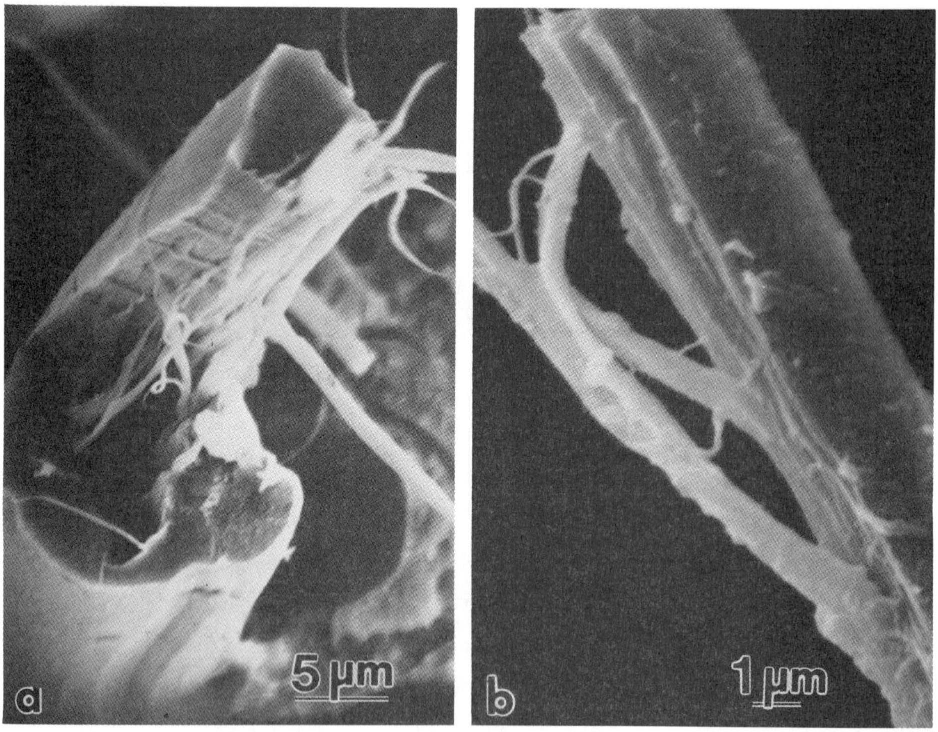

Figure 13. (a) The failed end of a human-made aramid fiber (PBO); (b) the failed end of one of the chitin fibers used to form the *Odontotaenius disjunctus* cuticle.

Figure 14. A "furry" chitin fiber, characteristic of microfibrils that have peeled away from the bundle to improve bonding to the matrix.

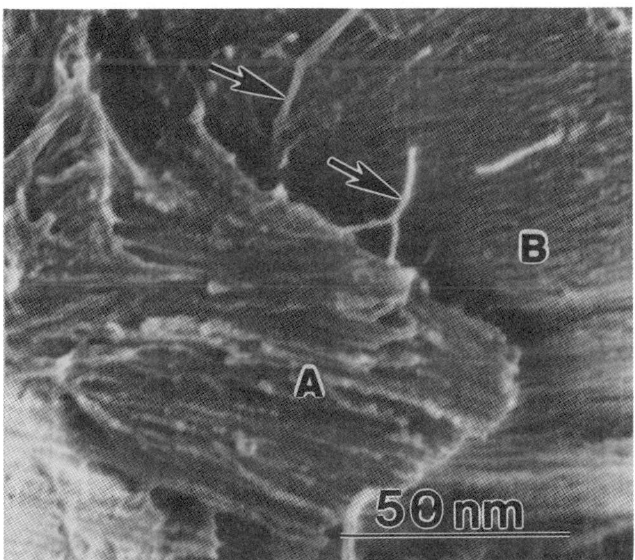

Figure 15. A high resolution SEM micrograph of the *Odontotaenius disjunctus* cuticle displays the fibrous matrix (a) and a chitin fiber (b). Note the microfibers between the fiber and the matrix (arrows).

bessbeetle cuticle possesses relatively good damage tolerance in the presence of a highly crosslinked, stiff matrix and a strong fiber/matrix interface, characteristics which normally result in a brittle system with poor damage tolerance. Examples of brittle synthetic systems which could benefit from mimicking nature's toughening mechanisms are ceramic-matrix, metal-matrix, and carbon-carbon composites, as well as highly cross-linked FRPC.

The bessbeetle cuticle is permeated by a number of holes (pore canals and dermal glands) that travel through the thickness of the cuticle and are used for nourishment and reconstruction. These holes, which are most apparent in sections of elytra, are produced *in-situ* during the formation of the exoskeleton.[11] As a result, the fibers remain continuous (or unbroken) in the vicinity of the holes by traveling around them (Figure 17). In contrast, holes are drilled or punched in man-made composites *after* fabrication, resulting in micro-cracking as well as additional stress concentrations at the broken fiber ends surrounding the hole. The insect therefore avoids, or at least minimizes, these undesired features with its novel design. Some microscopic observations of the dermal glands in bessbeetle cuticle revealed cracks that initiated away from the holes and then traveled around rather than through them (Figure 18).

3.0 MIMICKING UNIQUE DESIGNS USING SYNTHETIC MATERIALS

Of the many interesting and novel designs discovered in the bessbeetle cuticle, three were selected for mimicking using aerospace grade materials: the unsymmetrical layup, use of non-circular fibers, and molded or *in situ* formed holes.

3.1 UNSYMMETRICAL LAYUP

In this study, various mechanical properties of specimens having standard quasi-isotropic and single helicoid ply orientations were compared with those of specimens having the dual helicoid ply orientation observed in the bessbeetle cuticle. AS4/3501-6, a synthetic graphite fiber/epoxy resin material, was used for the study. The dual helicoid ply orientation of the bessbeetle cuticle is unsymmetric and unbalanced, while the standard quasi-isotropic and single helicoid orientations are both symmetric and

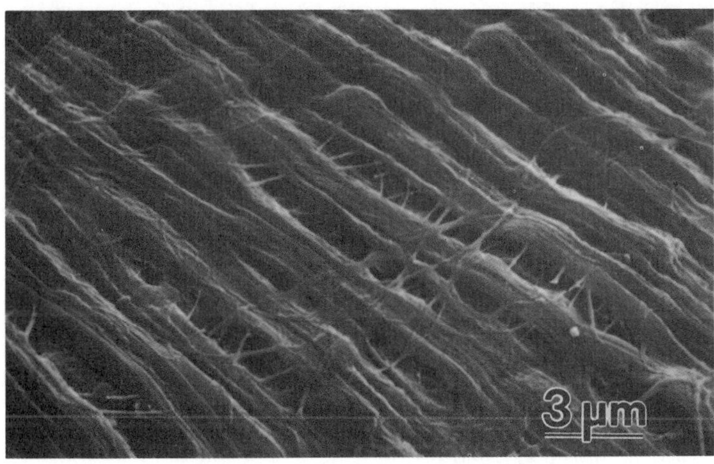

Figure 16. Micrograph displays the net-like appearance of microfiber bridging between adjacent fibers and plies.

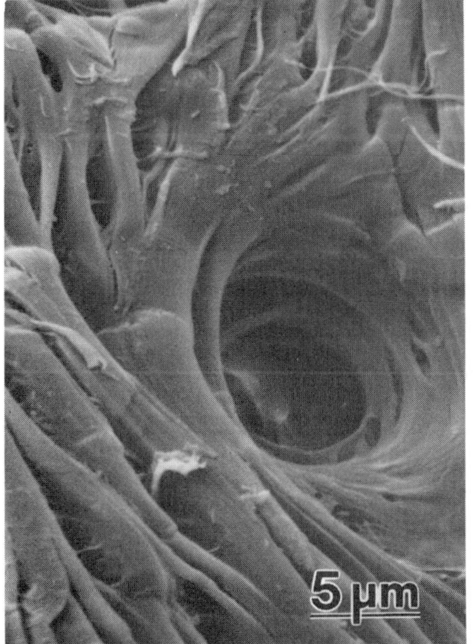

Figure 17. A dermal gland in the *Odontotaenius disjunctus* elytra. Note how the fibers maintain continuity in the presence of the hole by traveling around it.

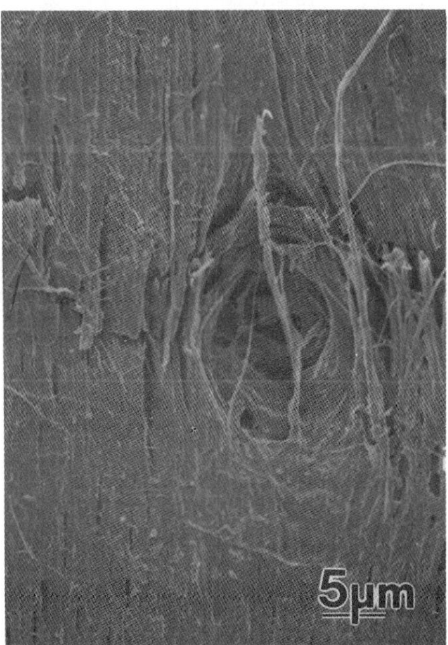

Figure 18. A crack in the *Odontotaenius disjunctus* elytra that has traveled around rather than through a dermal gland.

balanced. The specific stacking sequences used are as follows (note: both are 16 plies total, S represents the plane of symmetry):

Quasi-Isotropic (Symmetric): [0°/+45°/-45°/90°/0°/+45°/-45°/90°]S
Single Helicoid (Symmetric): [0°/-26°/-45°/-71°/+80°/+56°/+30°/+10°]S
Dual Helicoid (Unsymmetric): [0°/+85°/-26°/+61°/-45°/+36°/-71°/+9°/+80°/ -15°/+56°/35°/+30°/-63°/+10°/-86°]T

Theoretical analyses were performed with the GENLAM computer program which makes use of laminated plate theory for elastic properties and the Quadratic Interaction or Tsai-Wu failure criterion to predict strength. Experiments included longitudinal and transverse tension, compression, and flex tests as well as low velocity impact.

Due to the fact that the dual helicoid layup is highly unsymmetric and may create internal residual cure stresses that may effect the outcome of results, the COMPCAL computer program was used to estimate the residual stresses produced by each of the layups. The results from these predictions showed that the quasi-isotropic layup produced the lowest residual stresses with the unsymmetric dual helicoid surprisingly not far behind. The symmetric single helicoid layup actually produced the greatest residual stresses, however, none were great enough to effect the experimental results.

Theoretical and experimental values for elastic constants and strength are shown in Table-I and Table-II, respectively. All data is linearly extrapolated to 60% fiber volume content. Actual volume fractions varied from 59 - 63%.

Results in Table-I indicate good agreement between theory and experiment for elastic moduli and Poisson's ratio. The unsymmetric laminate shows similar moduli in both bending and inplane loading compared to the symmetric layups. In particular, the principal bending moduli are almost equal and very similar to the inplane moduli for the unsymmetric laminate, while the bending moduli of the symmetric laminates are quite different in the two principal directions and are also quite different than the inplane moduli. A better balance between bending and inplane elastic moduli could be obtained for the quasi-isotropic laminate by using a $[0°/90°/\pm45°]_S$ stacking sequence rather than the one investigated. However, even in this case the balance between

Table-I. Elastic Properties (60% Vf)

MODULUS, GPa (msi)						
	QUASI-ISOTROPIC		DUAL HELICOID		SINGLE HELICOID	
TEST	**Theory**	**Experiment**	**Theory**	**Experiment**	**Theory**	**Experiment**
Long. Tension	53.4 (7.75)	51.5 (7.48)	54.7 (7.94)	52.2 (7.57)	61.4 (8.91)	58.0 (8.42)
Long. Compr.	53.4 (7.75)	48.7 (7.07)	54.7 (7.94)	48.9 (7.09)	61.4 (8.91)	53.0 (7.68)
Trans. Tens.	53.4 (7.75)	49.0 (7.11)	52.5 (7.62)	50.7 (7.36)	47.3 (6.86)	42.5 (6.16)
Trans. Compr.	53.4 (7.75)	48.2 (6.99)	52.5 (7.62)	47.8 (6.94)	47.3 (6.86)	37.1 (5.38)
Long. Flex	70.3 (10.2)	59.9 (8.69)	60.8 (8.83)	53.7 (7.80)	67.5 (9.79)	65.1 (9.32)
Trans. Flex	37.3 (5.42)	33.3 (4.83)	59.7 (8.67)	52.0 (7.54)	35.1 (5.09)	28.4 (4.12)
POISSON'S RATIO						
x-y	0.31	0.30	0.30	0.29	0.34	0.34
y-x	0.31	0.29	0.31	0.28	0.34	0.28

Table-II. Ultimate Strengths (60% Vf)

STRENGTH, MPa (ksi)						
	QUASI-ISOTROPIC		DUAL HELICOID		SINGLE HELICOID	
TEST	**Theory**	**Experiment**	**Theory**	**Experiment**	**Theory**	**Experiment**
Long. Tension	608 (88.3)	617 (89.6)	615 (89.2)	615 (89.2)	764 (111)	527 (76.5)
Long. Compr.	660 (95.8)	555 (80.6)	613 (88.9)	443 (64.3)	850 (123)	569 (82.5)
Trans. Tens.	608 (88.3)	650 (94.3)	604 (87.6)	546 (79.2)	574 (83.2)	368 (53.4)
Trans. Compr.	660 (95.8)	505 (73.3)	637 (92.4)	441 (64.0)	693 (101)	357 (51.8)
Long. Flex	646 (93.8)	834 (121)	641 (93.0)	861 (125)		917 (133)
Trans. Flex	509 (73.9)	583 (84.6)	613 (89.2)	682 (99.0)		375 (54.3)

inplane and bending moduli would not be anticipated to be as good as in the unsymmetric laminate.

Comparison between theory and experiment for strength indicates much less correlation than found with elastic properties. This was anticipated for a number of reasons. The fracture process is very complex beyond first-ply failure. In addition, free-edge effects are present for both inplane and flexural loading. All of the specimens (whether they were loaded in tension, in compression, or in bending) displayed some delamination as part of the failure process. Delamination is not included in the failure criterion or in the

Table-III. Low-Velocity Impact Results

ORIENTATION>	SO	SH	DH (0 F)	DH (90 F)
INCIPIENT PARAMETERS				
Load, N (s.d.)	1571 (60)	2683 (206)	1822 (149)	1725 (45)
Energy, J (s.d.)	0.63 (0.04)	2.85 (0.65)	0.81 (0.11)	0.75 (0.05)
% ENERGY LOSS				
IMPACT ENERGY				
2 Joules	50	46	55	53
3 Joules	71	64	73	74
4 Joules	76	71	80	77
5 Joules	83	80	85	84
DAMAGE DIAMETER (cm)				
IMPACT ENERGY				
2 Joules	2.5	2.8	2.2	2.6
3 Joules	2.9	3.1	2.9	3
4 Joules	3	3.3	2.9	3.2
5 Joules	3	3.2	3.4	3.5

post first-ply failure process. Experimental data indicate that the inplane strength of the quasi-isotropic laminate under both tension and compression loading is higher than for the single or dual helicoid laminates. Care should be exercised, however, in making conclusions concerning comparison between the laminates for the case of compression loading, as such results may be highly influenced by the test method utilized in measuring compression strength. In addition, strain gage data from many of the compression specimens indicated some presence of buckling prior to failure. No indication of out-of-plane effects were observed for the dual helicoid specimens subjected to tension.

The impact results are shown in Table-III with the values in parentheses representing the standard deviation. The incipient parameters are indicative of the minimum load and respective energy required to initiate damage. In these terms, the unsymmetric specimens performed better than the quasi-isotropic specimens while the single helicoid layup possessed the highest incipient parameters indicating very good damage tolerance.

Percent energy loss represents the amount of applied energy absorbed by the specimen through heat, friction, vibration, and/or damage. Generally, the greater the percent energy loss, the greater the damage. Percent energy loss and damage area for all the layups in this study were similar.

This information indicates that the unsymmetric layup of the dual helicoid produces no detrimental effects with respect to low-velocity impact and in fact requires a greater load and higher energy to initiate damage compared to the quasi-isotropic layup. It should be noted, however, that the impact results cannot be considered conclusive since insufficient specimens were analyzed to obtain statistically correct data.

3.2 NON-CIRCULAR FIBERS

The influence of fiber cross-sectional shape and orientation on the properties of single- and multiple-ply laminates were theoretically analyzed. Finite-element analysis was used to examine ply stiffness and stress concentrations at the fiber/matrix interface where the orientation of elliptical fiber cross-sections were varied with respect to the ply surface.[12] Figure 19 displays the relationship between fiber shape and stress concentrations at the fiber/matrix interface when the one-fiber thick ply specimen is subjected to a transverse tensile load. The abscissa is the ratio of the fiber cross-sectional dimension in the y-direction to that in the z-direction (Figure 20); consequently, a value less than 1.00 represents an elliptical fiber cross-section with its major axis perpendicular to the ply surface, a value equal to 1.00 represents a circular fiber, and a value greater than 1.00 represents a fiber with its major axis parallel to the ply surface. The results indicate that a vertically-oriented elliptical fiber has a much lower stress concentration at its interface than the horizontally-oriented elliptical fiber or the circular fiber.

Figure 21 represents the extension of a specimen in the x-direction resulting from a transverse tensile load as a function of fiber axis ratio. Generally, it indicates that a ply with vertically-oriented elliptical fibers is more compliant than one with circular or horizontally-oriented elliptical fibers.

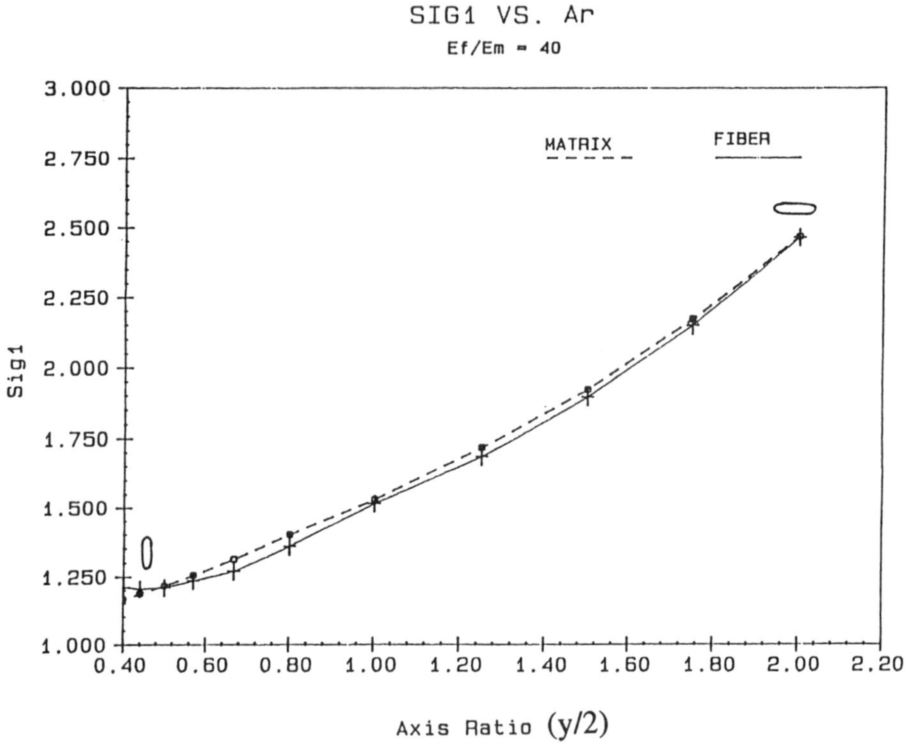

Figure 19. Relative stress concentrations at the fiber/matrix interface (sigma 1) results from a transverse tensile load as a function of fiber cross section geometry orientation.

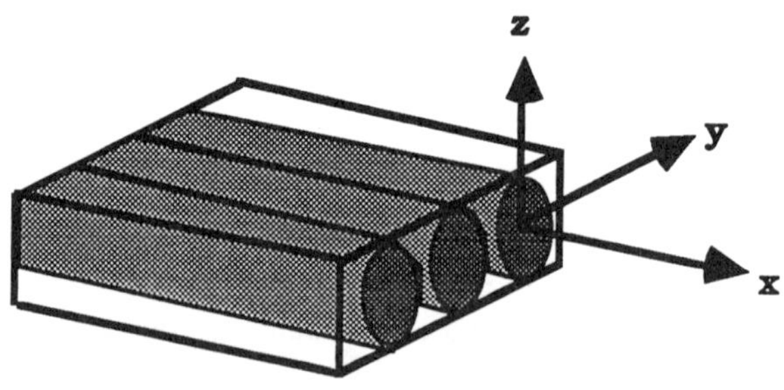

Figure 20. A schematic diagram displays the ordinate system used in the non-circular fiber analysis.

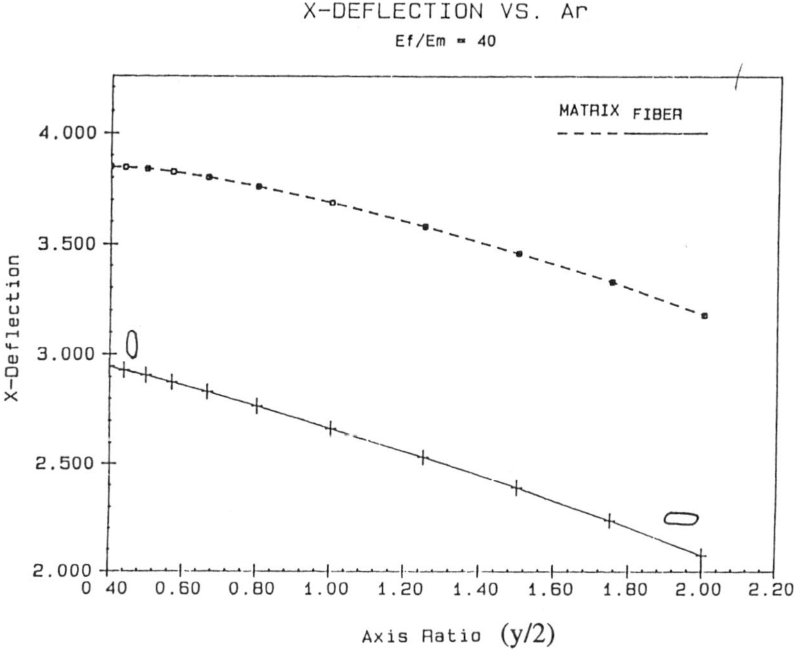

Figure 21. Relative extension of a specimen subjected to a transverse tensile load as a function of fiber cross sectional geometry orientation.

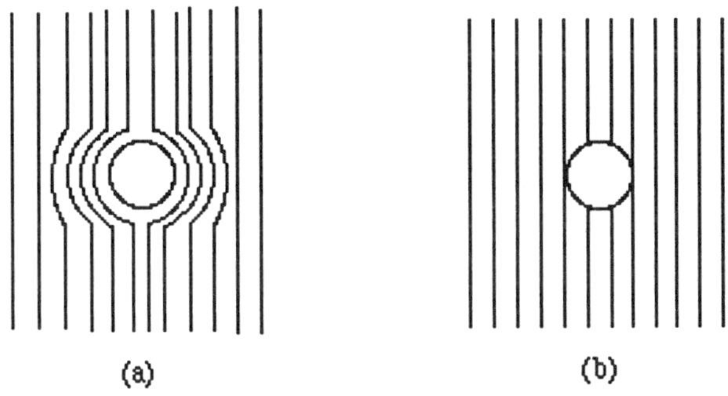

Figure 22. Schematic diagrams represent fiber orientation in the presence of a molded (a) and a drilled (b) hole.

3.3 MOLDED OR PREFORMED HOLES

The idea of using molded or preformed holes in man-made FRPC is not new,[13-17] although there has not been much development in this area. The appeal of molding holes, *in situ*, in composite parts lies in maintaining fiber continuity around the hole (Figure 22), a possible increase in the fiber volume fraction around the hole region, and a reduction in production costs.[13] These features result in decreased stress concentrations (and hence greater strengths), improved safety and part life, and overall savings.

To form the holes prior to processing, epoxy prepreg sheets were laid up on a steel plate with pins protruding through the bottom, thus creating holes in the laminate that could be preserved throughout processing. Tensile and compressive tests were run on drilled- and molded-hole specimens as well as unnotched specimens having $[\pm30°]_{4S}$ and $[0°/\pm45°/90°]_{2S}$ (quasi-isotropic) ply orientations. The $[\pm30°]_{4S}$ specimens were loaded continuously to failure while the quasi-isotropic specimens were loaded intermittently to evaluate damage initiation, propagation, and distribution. Specifically, the quasi-specimens were loaded to a predetermined level, unloaded, then viewed using X-ray and a dye penetrant. This process was continued until the specimen failed. Acoustic emission was also utilized in some cases to evaluate damage as a function of number of hits.

Fiber volume fractions of the drilled- and molded-hole regions were determined by acid digestion and showed almost a 4% increase for the molded holes. This results from not removing any material to form the molded holes. Also, micrographs of the hole cross-sections showed a much more dense packing of fibers around the molded holes.

The results of the $[\pm30°]_{4S}$ tension tests with the values standardized for a 60% fiber volume fraction are shown in Table-IV. The molded-hole specimens were significantly stronger than the drilled-hole specimens retaining 81% of the unnotched specimen strength. Observation of the failed specimens yielded variations in failure mode as well. The drilled hole specimens failed cleanly through the hole by shear and tensile failure with each specimen half containing half of the hole. The molded-hole specimens, however, exhibited a type of "hole pull-out" failure mode in which each broken half of the specimen contained a majority of the hole (Figure 23). The failure of the

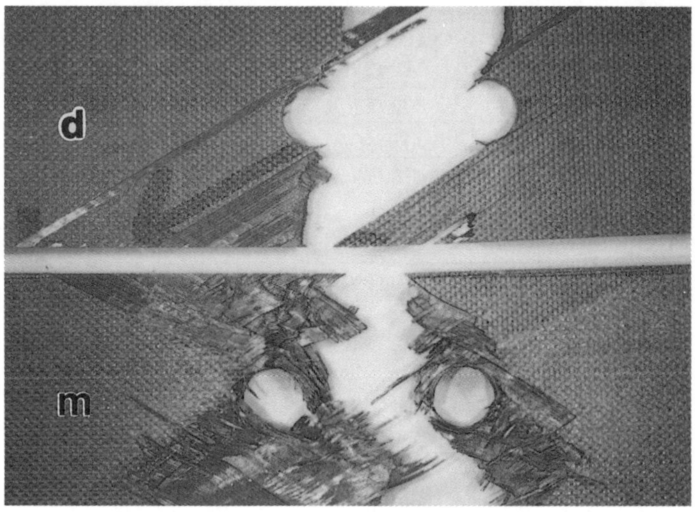

Figure 23. A failed $[\pm 30°]_{4S}$ drilled- (d) and molded- (m) hole specimen and their different failure characteristics. *(See also **Color Plate 4**.)*

molded-hole specimens was influenced by tension, shear, as well as delamination.

The ultimate strengths of the quasi-isotropic specimens are also shown in Table-IV. Again, a marked increase in strength is displayed by the molded-hole specimens compared to the drilled-hole specimens. Observation of the specimen X-radiographs taken at the different points during loading indicates that 90° matrix cracking was the first form of damage to occur in both specimens. As the load was increased the amount and distribution of matrix cracking also increased. It was also at these higher load levels that we began to notice a difference in the propagation of damage between the drilled- and molded-hole specimens. For the drilled-hole specimens the matrix cracks are concentrated across the ligament areas on either side of the hole, while for the molded-hole specimens the matrix cracks are more widely dispersed with a slight concentration occurring just above and below the hole (Figure 24). This seems to indicate that the areas of maximum stress and strain are occurring at the drilled hole and away from the molded hole illustrating the ability of the molded hole design to suppress stress concentrations and local strain.

Table-IV: Open-Hole Tension Test Results

ORIENTATION	HOLE TYPE	STRENGTH MPa (ksi)	STANDARD DEVIATION	STRENGTH RETENTION
[±30°]	Unnotched	592 (85.9)	46.6 (6.75)	
[±30°]	Drilled	343 (49.8)	15.0 (2.17)	58%
[±30°]	Preformed	478 (69.3)	21.7 (3.14)	81%
[0°/±45°/90°]	Unnotched	470 (68.2)	33.4 (4.84)	
[0°/±45°/90°]	Drilled	299 (43.3)	23.9 (3.46)	63%
[0°/±45°/90°]	Preformed	354 (51.3)	23.6 (3.42)	75%

Figure 24. An X-radiograph of a drilled- (d) and molded- (m) hole quasi-isotropic tension specimen at 60% and 65%, respectively, of the unnotched specimens' ultimate strength. Ninety-degree matrix cracks are not present across the molded hole.

There was no premature warning from the drilled-hole specimen as to its catastrophic failure while audible cracking could be heard from the molded-hole specimens just prior to failure suggesting improved damage tolerance. Acoustic emission results supported this claim in that the average number of hits sustained by the drilled-hole specimens was approximately 648 compared to 4270 hits for the molded-hole specimens.

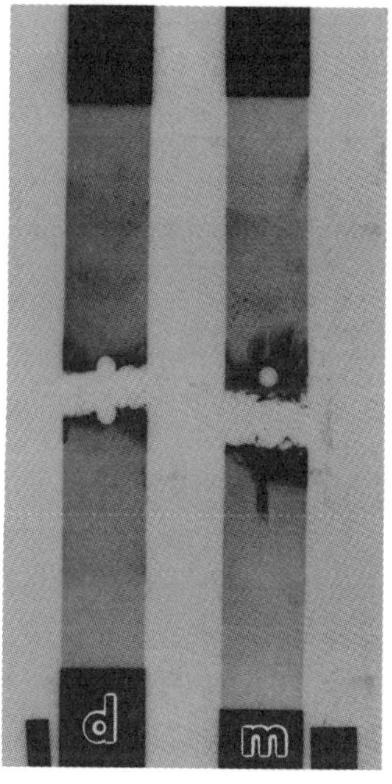

Figure 25. An X-radiograph of a failed drilled- (d) and molded- (m) hole quasi-isotropic tensile specimen. The latter specimen failed at a much higher load and accumulated more damage.

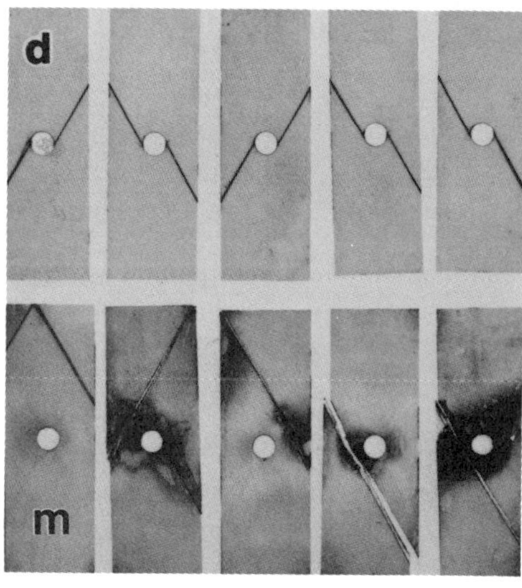

Figure 26. X-radiographs of failed drilled- (top row) and molded-hole [±30°] compression specimens.

The appearance of the failed quasi-isotropic specimens was similar to the [±30°] specimens in that the drilled-hole specimens failed straight across the hole and the molded-hole specimens produced a "hole pull-out" failure or in some cases even failed away from the hole. X-radiographs of the failed specimens indicate significantly more damage in the molded-hole specimens including extensive delaminations compared to the drilled-hole specimens (Figure 25) which further supports the claim of improved damage tolerance for the molded-hole design.

Results from the compression tests are shown in Table-V. Unnotched specimens were not run. Results of the [±30°] specimens show a significant increase in strength (50%) for the preformed-hole specimens compared to the drilled-hole specimens. The reason

Table-V. Open-Hole Compression Test Results

ORIENTATION	HOLE TYPE	STRENGTH MPa (ksi)	STANDARD DEVIATION	STRENGTH INCREASE
[±30°]	Drilled	241 (34.9)	3.66 (0.53)	
[±30°]	Formed	359 (52.1)	7.79 (1.13)	49%
[0°/±45°/90°]	Drilled	317 (45.9)	8.41 (1.22)	
[0°/±45°/90°]	Formed	348 (50.5)	92.4 (13.4)	10%

for this increase was revealed through failure analysis. While the drilled-hole specimens failed in shear along the 30° planes tangent to the hole, the preformed-hole specimens failed by delaminating around the hole region and shearing along the 30° planes away from the hole (Figure 26). This complex failure of the preformed-hole specimens provides increased strength.

Results from the quasi-isotropic specimens were not as clear, since significant scatter was evident in the preformed-hole specimen data. The average increase in strength for the preformed-hole specimens based on these results was small (10%) but is irrelevant due to the large scatter. Both specimen types failed in mixed-mode fashion, with compression and delamination providing most of the damage. X-radiographs of the failed specimens showed similar failure characteristics, with the preformed-hole specimens appearing to accumulate slightly more damage.

4.0 BIOMIMETIC APPLICATIONS TO SMART TECHNOLOGIES

An investigation of natural sensors to determine possible applications in the design of smart technologies is currently underway. The smart technologies include the areas of smart structures, smart skins, and smart processing. These technologies will be essential for the development of a smart aerospace vehicle (SAV) designed to sense or monitor its flight environment as well as its structural integrity. The SAV is proposed as a vehicle with arrays of sensors and computers, mimicking the role of the nervous system and brain respectively. This vehicle would be capable of monitoring in-flight loads, environmental conditions, hostile threats, and the health or well-being of the vehicle including detection and assessment of damage.[18,19] For this vehicle to become a reality, smart structures and smart skins must be developed.

Smart structures are defined as structures with incorporated sensors which monitor various parameters including: the health and flight-worthiness of the structure, the flight environment, loads during flight, and the extent and growth of damage. Theoretically, the introduction of a sensor into the structure would enable better monitoring of the part by detecting and evaluating damage unseen to the human eye, thereby improving safety, reliability, and maintainability, and resulting in less downtime of the aircraft. The direct monitoring of the structure's usage loads along with damage detection could also reduce the weight of the aircraft by avoiding over-design.[18,19]

The term smart skins is used to describe the placement of electromagnetic sensors just below the surface of the structure. The sensors act as receivers and transmitters to monitor hostile threats, provide communication links, and indicate appropriate counter-measures. The distribution of these sensors throughout the vehicle instead of in cutouts in the skin or on bulky equipment racks in the fuselage could improve radar coverage, warning and direction finding, target recognition and identification, and communication.[18]

Smart processing employs the use of sensors and computers to monitor and control various parameters during the fabrication of a part. These parameters include temperature of the part, evolution of volatiles, viscosity, compaction and strain, and the detection of processing-induced damage such as microcracks. A combination of sensors with an expert system could more closely control or optimize fabrication, resulting in reduced processing time and increased reliability.

In order to utilize the above mentioned sensors, several problems must first be addressed; the first is the selection of the appropriate sensors and materials. Some of the sensors and materials which are being investigated are fiber-optics, acoustic waveguides, acoustic emission, piezoelectric, shape memory alloys, and electro-rheological fluids.[18, 20] The sensors and materials desired are those which can detect more than one parameter and survive from fabrication through the life of the part.

Another critical problem facing smart structures and skins is the placement of sensors, both grossly in the structure and molecularly in the materials. In composite materials the sensors must survive the conditions during fabrication (temperature and pressure), have minimal effects on the mechanical properties of the part, and not be impaired or damaged by environmental conditions during the life of the part. A final issue will be the interconnection of, and collection of data from the sensors, data reduction, and output of information to the appropriate systems.

To solve some of these problems, researchers can look to materials that are inherently smart, those made by nature. Biological systems are composed of innately smart materials and structures which are designed to collect and process information in real-time on the animals' well-being or health. For example pain, thirst, hunger, temperature, position of body parts, rate and direction of body movement , blood pressure, pH balance, and levels of oxygen and carbon dioxide are just some of the information that is constantly being monitored. Simultaneously, the animal processes information about its environment through visual, auditory, tactile, olfactory, and gustatory (taste) senses.[21]

The control of smart materials in living organisms is through the nervous system which, along with the endocrine system, dictates most of the functions of the animal's body. The nervous system controls rapid activities such as muscular contraction, visceral (instinctive) activities, and the rate of secretion from some endocrine glands. The endocrine system on the other hand regulates metabolic rates such as transport of substances through cell membranes, rate of chemical reactions, and cellular metabolisms such as growth and secretion.[22] The nervous system in animals can easily be compared to the sensors envisioned for the SAV.

One area of study in the nervous system which could help in the design of advanced systems is the examination of unique sensors in living organisms and the method by which they function. In the animal kingdom there are five basic types of sensory receptors. They include: (i) chemoreceptors, which detect oxygen content in blood, osmolality of body fluids, carbon dioxide concentrations, and olfaction and taste;
(ii) thermoreceptors, which detect changes in temperature; (iii) mechanoreceptors, which detect mechanical deformation of the receptor or the surrounding area;

(iv) electromagnetic receptors, which respond to light; and (v) nociceptors or pain receptors, which detect damage to the surrounding tissue. Each of these receptors has a different structure which allows it to react to specific stimuli. Each sensor translates its information by electrical inputs or action potentials via the afferent nerves to the brain. The receptors work essentially as transducers, transferring information, be it electromagnetic radiation or mechanical vibrations, from one form of energy or code to another.

In the animal body the somatic senses are sensors which collect information on the well-being of the organism. This would be analogous to sensors on an aircraft sensing the health of the aircraft including the detection of damage. The somatic senses are made up of three different types of sensors (i) mechanoreceptors, (ii) thermoreceptors, and (iii) nociceptors. The somatic senses can also be broken down into four different sensations depending on where the sensation is located. These include: (i) exteroceptive sensations (from the surface of the body), (ii) proprioceptive sensations (physical state of the body including position), (iii) visceral sensations (internal organs), and (iv) deep sensations (internal tissues such as bone).[22] Some of the specific receptors found in the above groups include free nerve endings, Merkel's disc, tactile hair, Pacinian corpuscle, Meissner's corpuscle, Krause's corpuscle, Ruffini's end-organ, Golgi tendon apparatus, and muscle spindles.

The mechanoreceptors are composed of tactile senses, including touch, pressure, and vibration; and kinesthetic senses which determine the position and rate of body movement. The tactile receptors are rapidly-adapting receptors which react quickly to a stimulus but rapidly become equalized and return to a resting state. Some of the receptors associated with tactile sensations include free nerve endings, Meissner's corpuscle, Merkel's disc, hair end-organ, Ruffini's end-organs, and Pacinian corpuscles.[21-23]

Aside from the sensors of the skin, there are many other specialized sensors in animals that are unique and may aid in the design of synthetic sensors for smart structures and skins. For example, some snakes sense infrared radiation,[24,25] bats and certain fish use echo-location to travel or find prey,[21,24] several insects sense ultraviolet radiation,[24] and some fish use electrical location for direction and to detect prey.[26]

One of the characteristics desired in smart skins is the detection of various wavelengths of electromagnetic radiation from ultraviolet to infrared. In nature, UV wavelengths are detected by bees and some ants,[24] while some snakes use IR detection to hunt for prey. The infrared sensors in the Crotalinae (pit vipers) and Boidae (boid) snakes are in the form of pit organs located in deep cavities in the head. The pit vipers, including cottonmouth, copperhead, and rattlesnakes, have pit openings on the side of the head below and in front of the eyes.[27,28] The boid snakes including pythons and boas can have as many as 13 pairs of pits or heat-sensitive scales bordering the mouth. The pit organ is thought to have evolved from the somatic sensory system. In the Crotalinae snakes, the most sensitive of the two groups, the pit organ, measures about 1 mm to 5 mm in diameter and is located in the soft tissue and bone of the head. It is covered by (approximately) a 30 square mm by 15 μm thick membrane which has about 7,000 thermosensitive endings.[28] Aside from the pit organ, the snake also has a unique brain structure that is responsible for processing the IR information.

Another possible area of interest for smart technologies could be the study of sonar in bats. Bats use high-frequency sound waves to detect prey and avoid obstacles. Some bats are even capable of catching fish, relying on their sonar to track prey through air and water.[25] The most interesting aspect of the bat's use of sonar is its ability to detect prey and avoid objects in spite of noise or jamming. A study to test the discriminatory power of the bats used continuous broad-band noise to disorient the bats.[25] They found that the bats could still evade insect nets and wires 1 mm in diameter. In tomorrow's aerospace smart skins, interference from other equipment and jamming are key issues which will need to be addressed.

The study of biological concepts to aid in the design of smart structures has already been documented by a group from the Florida Institute of Technology.[29] In this study a neuro-muscular unit, for example a leg or arm, was used as a model to design and build airfoil control surfaces as well as artificial joints. Similarly, the study of the somatic system could provide insight to the development of sensors which could aid in health monitoring and damage detection; while the investigation of specific sensors such as infrared, ultraviolet, and electric could help in the design of smart skins.

5.0 CONCLUSIONS

It is evident that nature offers a vast supply of design ideas, and many of these designs can be incorporated into man-made structures to improve performance and versatility. Initial studies using selected novel designs indicate that improvements are definitely possible; however, difficulties arise in biomimetics from a lack of understanding of natural materials and problems in reproducing or mimicking their designs with synthetic materials using conventional fabrication techniques. Also, it may not be just *one* of the many design concepts in an organism that allow the function of that design to be successful, but all of the designs, and thus the system, acting in unison. Additional efforts must be made at understanding the chemistry, processing, and purpose associated with each design in order to benefit from it. Furthermore, one must not forget the precision and control Mother Nature employs in forming the natural designs, the levels of structure present, and their interaction.

6.0 ACKNOWLEDGEMENTS

The authors would like to gratefully acknowledge the Air Force Office of Scientific Research (AFOSR) and Wright Laboratories (WL) for funding of the task and Jim Whitney, Dave Anderson, Ron Cornwell, Jim Lute, John Sawvel, Eric Nelson, Joe Saliba, Peter Sjoblom, and many others for playing various supporting roles in the activities of this task.

7.0 REFERENCES

1. H. R. Hepburn and A. Ball, "On the Structure and Mechanical Properties of Beetle Shells," *J. Mater. Sci.* **8**, 618-623 (1973).

2. N. F. Hadley, "The Arthropod Cuticle," *Sci. Amer.*, 104-120 (1986).

3. *Structural Biomaterials,* J. F. V. Vincent (Princeton University Press, Princeton, 1990).

4. R. C. Schiavone and S. L. Gunderson, "The Components and Structure of Insect Exloskeleton as Compared to Man-Made Advanced Composites," in *Proceedings of the American Society for Composites, Fourth Technical Conference*, (Technomic, Pittsburgh, 1989) pp. 876-885.

5. S. L. Gunderson and R. C. Schiavone, "Structural Composites; Natural," in *The International Encyclopedia of Composites*, Lee C (ed.) (VCH Publishers, New York) (in press).

6. A. S. Crasto, et. al, "Improved Technology For Advanced Composite Materials," *Vol. III, AFWAL-TR-88-4241* (1991).

7. F. Guild, B. Harris, and A. Atkins, "Cracking in Layered Composites," *J. Mater. Sci.*, **13**, 2295-2299 (1978).

8. B. Zelazny and A. C. Neville, "Quantitative Studies on Fibril Orientation in Beetle Endocuticle," *J. Insect Physiol.*, **18**, 2095-2121 (1972).

9. H. R. Hepburn and I. Joffe, "On the Mechanical Properties of Insect Exoskeletons," in *The Insect Integument*, H. R. Hepburn (ed.) (Elsevier, Amsterdam, 1976) pp. 207-235.

10. G. Fraenkel and K. M. Rudall, "A Study of the Physical and Chemical Properties of the Insect Cuticle," *Proc. R. Soc. Lond.*, B, **129**, 1-35 (1947).

11. S. L. Gunderson and R. C. Schiavone, "The Insect Exoskeleton: A Natural Structural Composite," *JOM*, **41** [11] 80-82 (1989).

12. J. E. Saliba, R. C. Schiavone, S. L. Gunderson, and D. G. Taylor, "Mechanics of Natural Composites," in *Materials Synthesis Based on Biological Processes*, Proc. MRS Symp., M. Alper, P. Calvert, R. Frankel, P. Rieke, and D. A. Tirrell (eds.) Vol. **218** (Materials Research Society, Pittsburgh, 1991) pp. 215-220.

13. N. O. Brink, "Formed Fastener Holes in Composite Structures," in *Proceedings of the 25th National SAMPE Symposium and Exhibition*, 223-232 (1980).

14. L.-W. Chang, S.-S. Yau, and T.-W. Chou, "Notched Strength of Woven Fabric Composites with Moulded-In Holes," *Composites*, **18** [3] 233 (1987).

15. S.-S. Yau and T.-W. Chou, "Strength of Woven-Fabric Composites with Drilled and Molded Holes," in *Composite Materials: Testing and Design*, J. D. Whitcomb (ed.) (ASTM, Philadelphia, 1988) pp. 423-437.

16. M. N. Ghasemi Nejhad and T.-W. Chou, "Compression Behavior of Woven Carbon Fibre-Reinforced Epoxy Composites with Moulded-In and Drilled Holes," *Composites*, **21** [1] 33 (1990).

17. M. N. Ghasemi Nejhad and T.-W. Chou, "A Model for the Prediction of Compressive Strength Reduction of Composite Laminates with Moulded-In Holes," *J. Comp.*, **24** [3] 236 (1990).

18. C. J. Mazur, G. P. Sendeckyj, and D. M. Stevens, "Air Force Smart Structures/Skins Program Overview," in *Fiber Optic Smart Structures and Skins, SPIE, Vol. 986,* 19 (1988).

19. G. P. Sendeckyj and C. A. Paul, "Some Smart Structures Concepts," in *Fiber Optic Smart Structures and Skins II, SPIE,* Vol. 1170 (1989).

20. P. Donaldson, "Will Aircraft Feel Pain?" *Aero. Comp. and Mats.*, **3** [1] 21 (1991).

21. *Animal Physiology: Principles and Adaptations*, M. S. Gordon, 2nd ed. (Macmillan Publishing Co., Inc., New York, 1972).

22. *Textbook of Medical Physiology*, M. D. Guyton and C. Arthur (W. B. Saunders Co., Philladelphia, 1976).

23. *The Senses of Man*, J. S. Wilentz (Thomas Crowell Co., New York, 1968).

24. *The Senses of Animals*, E. T. Burtt (Wykeham Publications, London, 1974).

25. D. R. Griffin, "More About Bat Radar," *Sci. Amer.*, **199**, 40-47 (1958).

26. H. W. Lissman, "Electric Location By Fishes," *Sci. Amer.*, **204**, 15-25 (1963).

27. J. F. Harris, "Snake Infrared Receptors: Thermal or Photochemical Mechanisms?" *Science*, **172**, 1252-1253 (1971).

28. E. Newman and P. Hartline, "The Infrared Vision of Snakes," in *The Mind's Eye* (Scientific American Inc., Washington, DC, 1986).

29. M. H. Thursby, et al., "Smart Structures Incorporating Artificial Neural Networks, Fiber-Optic Sensors, and Solid-State Actuators," in *Fiber Optic Smart Structures and Skins II, SPIE,* Vol. 1170, 316 (1989).

STRUCTURE AND FUNCTION OF MAGNETOSOMES IN MAGNETOTACTIC BACTERIA

Richard B. Frankel* and Dennis A. Bazylinski#

*Department of Physics, California Polytechnic State University
San Luis Obispo, California 93407, USA
#Marine Science Center, Northeastern University
East Point, Nahant, Massachusetts 01908, USA

Magnetotactic bacteria contain magnetosomes, which are mineral particles enclosed by membranes. The particles are ferrimagnetic magnetite, ferrimagnetic greigite, or greigite and non-magnetic pyrite. The particles constitute an elegant biomagnetic compass that orients the cell along the geomagnetic field lines as it swims. This paper discusses the structures of these particles and their possible formation mechanisms.

1.0 INTRODUCTION

A number of species of motile, aquatic bacteria are able to orient and navigate along geomagnetic lines.[1-3] This behavior, known as magnetotaxis, is based on the fact that each magnetotactic bacterium is a swimming, permanent magnetic dipole, that is, a motile bio-magnetic compass.[2] The permanent magnetic dipole moment of each magnetotactic cell is due to intracellular, membrane-bound, permanent single-magnetic-domain-sized inorganic particles known as magnetosomes, which are, in most cases, arranged in chains.[1,3,4,5] The biomineralization process, involving the composition, size, position, orientation, and even morphology of the particles, is highly controlled by the bacteria.[6] Moreover the magnetosome chain is a hierarchical structure that is a masterpiece of permanent magnet engineering.

The magnetosomes of most of the magnetotactic bacteria that have been studied to date contain particles of magnetite, Fe_3O_4. An example is shown in Figure 1. This organism, designated strain MV-4, is an unidentified magnetotactic bacterium from a marine marsh that has recently been isolated and grown in pure culture. It contains a chain of 16 uniformly sized and shaped magnetite particles, each about 60 nm along the chain axis. Magnetite was identified as the mineral form by electron diffraction measure-

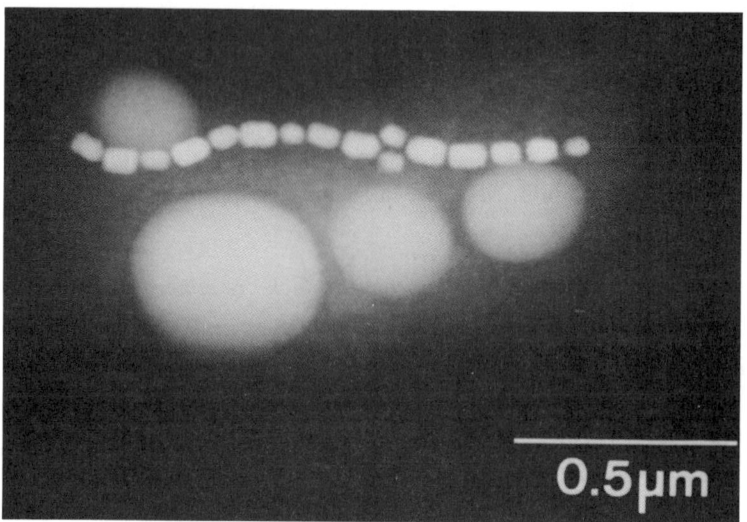

Figure 1. TEM/BF image of the marine, magnetotactic bacterium, strain MV-4. The chain of particles in the cell are the magnetite-containing magnetosomes. The globular structures contain polyphosphate or sulfur (see Figure 2).

ments on particles in whole cells.

In addition to the magnetosomes, cells of strain MV-4 contain other electron-dense structures, which also consist of partially inorganic materials. These are shown in Figure 2 where elemental maps of Fe, O, S, and P are shown for a cell of strain MV-4. These maps were produced in a scanning transmission electron microscope, fitted with a fluorescent X-ray detector. As the electron beam was rastered over the specimen, the instantaneous X-ray intensities of several selected elements were determined and recorded. It can be seen that Fe and O density correlate with the positions of the particles in the cell, which is consistent with their identification as magnetite. P and O map with the large, globular, electron-dense structures in the cell. These are presumably polyphosphate granules, which occur in many types of bacteria. Finally, two globular concentrations of S are also seen in the cell. These could be elemental sulfur globules which are obscured by the large polyphosphate granules in the electron micrograph. Thus the cell produced at least three spacially segregated inorganic products.

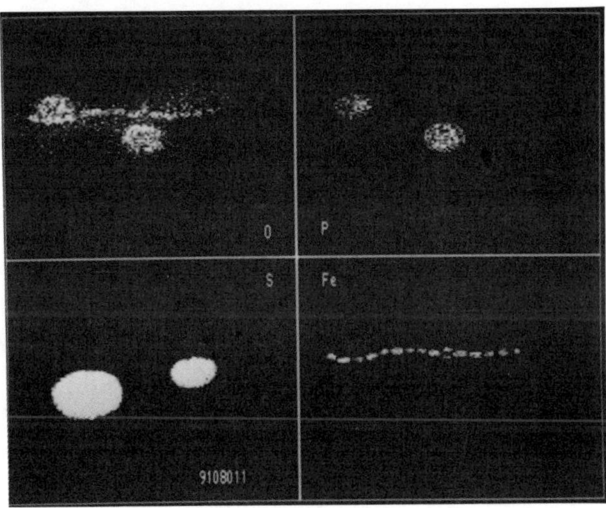

Figure 2. Elemental X-ray density maps of the organism shown in Figure 1: top left, 0; top right, P; bottom left, S; bottom right, Fe X-ray images.

The consistency of the magnetosome particle shape in strain MV-4 suggests a uniform particle morphology. This has recently been determined to be cubo-octahedral with a small elongation along the [211] direction.[7] In fact, narrow size distributions and species-specific morphologies are a characteristic feature of the magnetite biomineralization process in all magnetotactic bacteria.[6] These qualities are not characteristic of inorganically (chemically) produced magnetite, nor of extracellular magnetite produced by dissimilatory iron-reducing bacteria.[8] The latter type of indirect magnetite production results from chemical modifications of the extracellular environment by the organisms themselves. In this case the magnetite results from a chemical reaction between extracellular ferric oxyhydroxide and soluble ferrous ions which the cells secrete as they reduce ferric iron. The resulting magnetite particles have a wide size distribution and no organic component. This type of biomineralization has been termed "biologically-induced mineralization,"[9] in contrast to "biologically-controlled mineralization" in which the inorganic particles are produced within organic matrices.[9,10]

A number of idealized morphologies of bacterial magnetite in magnetotactic bacteria have been characterized[6] some of these are shown in Figure 3. The cubo-octahedral

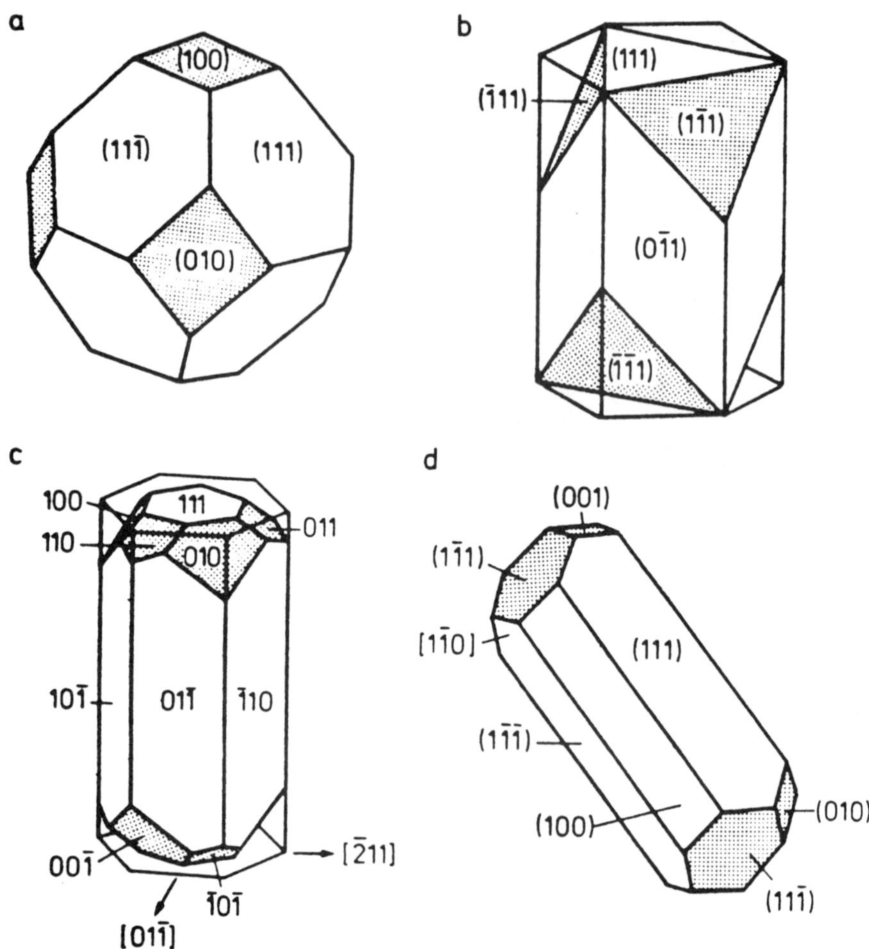

Figure 3. Idealized crystal morphologies of bacterial magnetite: (a) cubo-octahedron; (b) and (c) hexagonal prisms; (d) elongated cubo-octahedron (after Ref. 6).

morphology (Figure 3a) preserves the symmetry of the face-centered cubic, spinel crystal structure, and may be considered an equilibrium growth form. The other morphologies (Figures. 3b-d) all represent departures from the equilibrium form, presumably due to acceleration or deceleration of the growth of certain crystal faces,

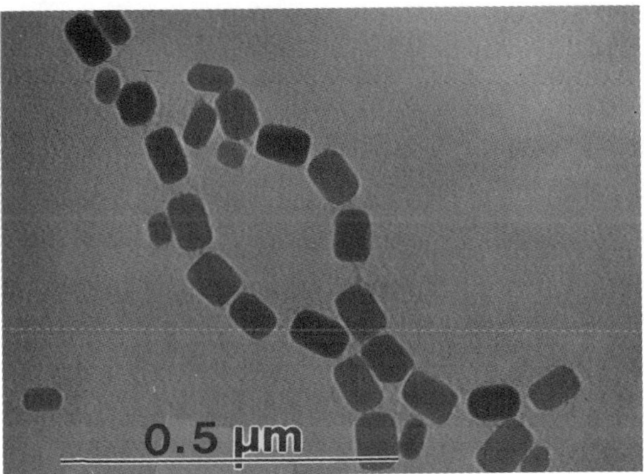

Figure 4. TEM/BF image of magnetite particles separated from cells of strain MV-1 showing the magnetosome membrane in this organism. Magnification: 100,000 x.

possibly resulting from the differential binding of macromolecules to those faces.

It is clear that magnetotactic bacteria exercise great control over the biomineralization process. How they do this is unclear but is currently under study. In several species of magnetotactic bacteria, the magnetite particles are enveloped in what appears to be a membrane vesicle. In *Aquaspirillum magnetotacticum*, the vesicle is a unit membrane (i.e., a lipid bilayer) consisting of phospholipids and numerous proteins, which is apparently not contiguous with the cytoplasmic membrane.[11] The magnetosome membrane is presumably the structural entity that anchors the mineral particle at a particular location in the cell, as well as the locus of control over the size and morphology of the particle. A magnetosome membrane has also recently been found in cells of the facultatively anaerobic, marine, magnetotactic bacterium, strain MV-1 (Figure 4). Clearly, more information about these membranes and their associated proteins in different magnetotactic bacteria would be highly desirable in elucidating the biomineralization process.

2.0 MAGNETIC STRUCTURE OF THE MAGNETOSOME CHAIN

The hierarchical structure of the magnetosome chain is significant when one considers its magnetic properties.[2] Firstly, consider the size of the individual particles. Large particles of any magnetic material, including magnetite, lower their magnetostatic energy by forming magnetic domains, thus reducing the remnant magnetic moment of the particle. Magnetic domains are regions of uniform magnetization and are separated from each other in the particle by transition regions known as domain walls. In the domain walls, the direction of magnetization changes smoothly from that of one domain to the other. The width of the domain walls in a particle is determined by the fundamental magnetic properties of the material including magnetic exchange and anisotropy energies, and is hence a constant for any magnetic material. Thus when the particle dimensions become comparable with the domain wall width, domains cannot form and the particle is forced to remain a single magnetic domain with uniform, maximum magnetization. For magnetite, this is 92 emu/g. Calculations by Butler and Banerjee[12] give 76 nm as the upper limit for the single-magnetic-domain-size range for equidimensional particles of magnetite. Because of shape anisotropy, the single-magnetic-domain volume increases with axial ratio for non-equidimensional particles. As a rule of thumb, magnetite particles with long dimensions of the order of 120 nm or less are single magnetic domains.

The thermal stability of the magnetization in single-magnetic-domain particles is determined by the particle volume. The magnetization is oriented along an energetically favorable direction in the particle known as an easy magnetic axis, which, for magnetite above the so-called Verwey transition at 118 K, is parallel to a <111> direction. There are several equivalent <111> directions in the lattice and thermal energy can spontaneously excite transitions of the magnetization over the intervening hard magnetic directions, or energy barriers due to magnetic anisotropy. This behavior, known as superparamagnetism, results in a time-averaged loss of remnant magnetization in an ensemble of particles. For single-magnetic-domain particles above a certain volume, the transition rate of the magnetization will be negligible and the particles will retain a permanent magnetization. For magnetite at 300 K, particles with dimensions greater than or equal to about 35 nm will be permanently magnetized. Thus, magnetite particles with long dimensions between about 35 and 120 nm are permanent, single magnetic domains at ambient temperature. The magnetite particles produced by magnetotactic bacteria are typically within this size range. Thus the bacteria are not only producing magnetic, mineral particles, they are producing permanent, single-magnetic-domain-sized particles of that mineral.

When the particles are organized into chains, as they are in magnetotactic bacteria, the magnetic interactions between them cause their magnetic dipole moments to orient parallel to each other along the chain direction. The total magnetic dipole moment of the chain is thus the sum of the moments of the individual particles. By organizing the particles into chains, a bacterium is essentially constructing a permanent magnetic dipole which is sufficiently large to orient the cell in the geomagnetic field in water as it swims. That is, the magnetic energy of the dipole in the geomagnetic field is about a factor of 10 or greater than thermal energy at ambient temperature, which results in an 80% or better average projection of the moment on the field direction.

The cellular magnetic dipole can have two possible orientations with respect to the flagellum, which makes the organisms either North-seeking or South-seeking in the geomagnetic field. Organisms with the former and latter polarity predominate in the Northern and Southern hemispheres, respectively.[1-3] The advantage of magnetotaxis is presumably increased efficiency in finding and maintaining preferred position in redox and/or oxygen gradients.

We might note that permanent magnet manufacturers have been using biomimickry without realizing it. The basic strategy in the manufacture of permanent magnets from different materials, including alnico, samarium-cobalt, neodymium-iron-boron, and others, is to first produce permanent, single-magnetic-domain-sized particles of the material and then to press the particles together and sinter them to form the finished magnet. The magnet is then magnetized in an intense magnetic field. The bacteria have engineered a more elegant solution by forming chains of single magnetic domains which spontaneously produce the maximum magnetization.

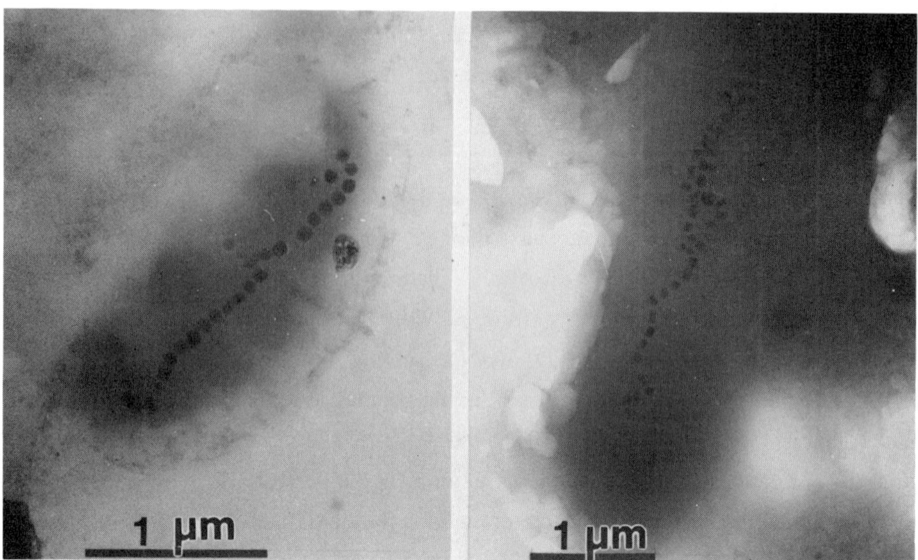

Figure 5. TEM/BF images of two magnetotactic rod-shaped bacteria collected from sulfidic aquatic habitats. Bars: 1 micron.

3.0 IRON SULFIDE PARTICLES IN MAGNETOTACTIC BACTERIA

Molecular oxygen is apparently required for cells of *A. magnetotacticum* to produce intracellular magnetite,[13] and this has led to the assumption that magnetite formation by magnetotactic bacteria is confined to surface sediments in aquatic habitats with microaerobic conditions.[14] However, morphologically diverse forms of magnetotactic bacteria are also common in reducing sediments and waters that contain high concentrations of hydrogen sulfide (H_2S) and probably no free oxygen.[15] Such conditions occur in coastal estuarine environments, salt marsh pools and certain shallow anaerobic basins.

Sulfide-rich sediments and water collected from both the west coast and east coast of the United States (Morro Bay, California and Woods Hole, Massachusetts, respectively) contained large numbers of rod-shaped magnetotactic bacteria. These were separated

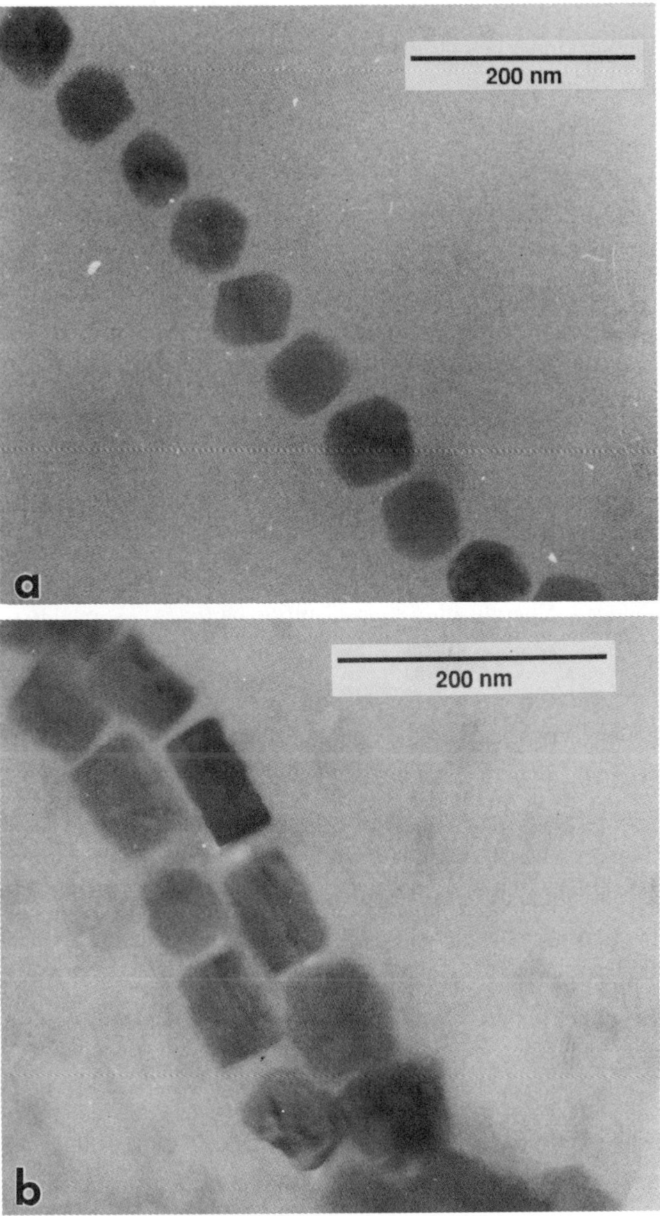

Figure 6. Magnified view of intracellular greigite, Fe_3S_4, particles in the cells shown in Figure 5, showing (a) cubo-octahedral and (b) rectangular prismatic morphologies.

magnetically and examined by electron microscopy. At least two types of rod-shaped organisms were found, a smaller one (ca. 2.5 x 1.3 mm) and a larger one (ca. 3 mm x 2 mm), containing cubo-octahedral and rectangular prismatic electron dense particles, respectively. These are shown in Figures 5 and 6. X-ray elemental mapping by scanning transmission electron microscopy revealed that the particles were composed of iron and sulfur, not iron and oxygen (Figure 7).[15] Identification of the iron-sulfur mineral phase by indexing isolated single crystal electron diffraction patterns revealed that both the smaller and larger rods contained particles of greigite, Fe_3S_4.[16,17] This mineral has a face-centered spinel structure and is isostructural with magnetite. Study of the particles by high resolution transmission electron microscopy revealed two distinct, idealized particle morphologies which are shown in Figure 8.[17] Thus greigite formation in these bacteria parallels magnetite formation in the microaerophilic bacteria described above, with narrow size distributions and species-specific morphologies, implying a high degree of control over the biomineralization process. Uncontrolled, biologically-induced mineralization of iron sulfide particles is known to occur in some dissimilatory sulfate-reducing bacteria.[18]

Greigite is ferrimagnetic at ambient temperature. Its saturation magnetization is approximately 30 emu/g, about one third that of magnetite. Thus it is less "efficient" than magnetite as a permanent magnet material. Nevertheless, it is perfectly adequate as the basis of the magnetotactic response in those bacteria which produce it. Many of the basic magnetic properties of griegite have not been determined so the single-magnetic domain size range is not known, but on the basis of the magnetization and Curie temperature we may infer that it is approximately the same as magnetite. Thus chains of magnetosomes containing greigite would function analogously to chains of magnetosomes containing magnetite.

Living in the same sulfidic habitats on the west and east coasts of the United States as the greigite-containing magnetotactic rods described above is an unusual multicellular, magnetotactic prokaryote (Figure 9) that is referred to casually as "the mulberry" because of its appearance in the light microscope.[19] A similar organism occurs in Brazil,[20] and probably in similar habitats in other parts of the world. The intact organism contains 7 to 20 individual cells, each of which has flagella on one side of the cell, and contains about 10 electron-dense particles arranged in chains Figure 10. There is

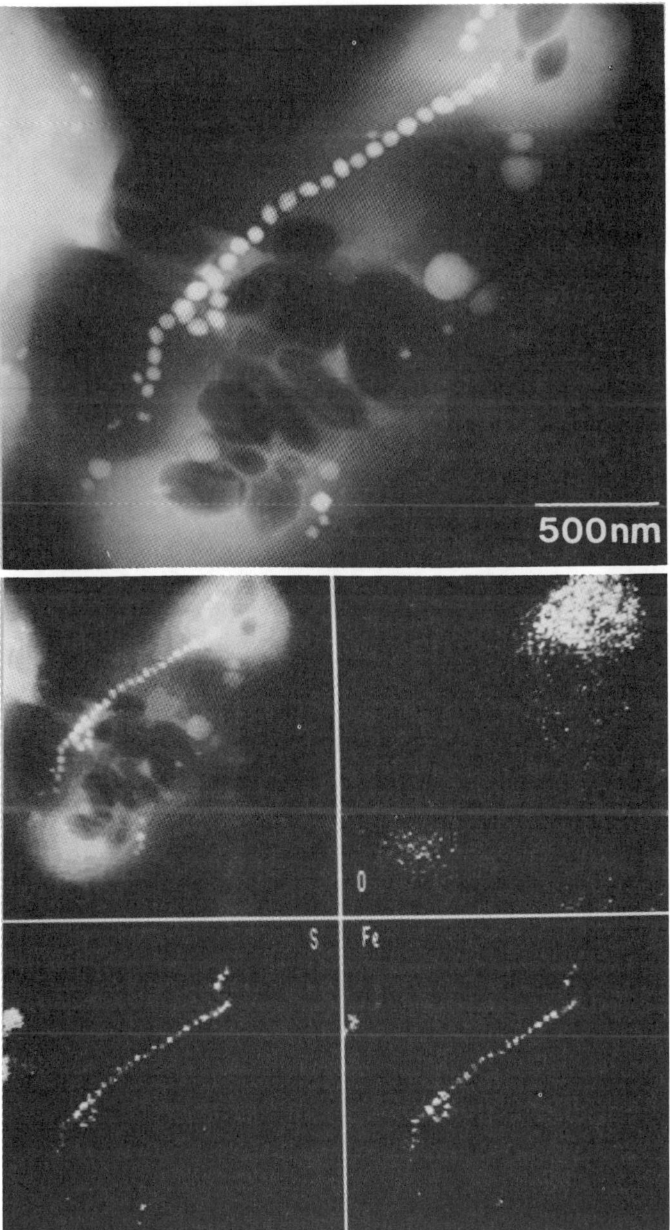

Figure 7. Elemental X-ray density maps for a magnetotactic rod-shaped bacterium with cubo-octahedral particles. Upper panel: STEM/BF image. Lower panel: upper left, STEM image; upper right, O; bottom left, S; bottom right, Fe. The correlation of Fe and S density with particle position shows that the particles contain Fe and S but not O. Similar results have been reported for cells containing rectangular prismatic particles.

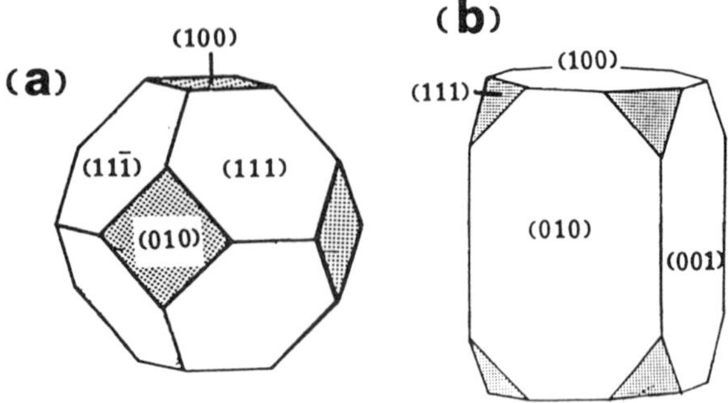

Figure 8. Idealized morphologies of greigite crystals formed in magnetotactic bacteria: (a) cubo-octahedron; (b) elongated, truncated cube. (After Ref. 17.)

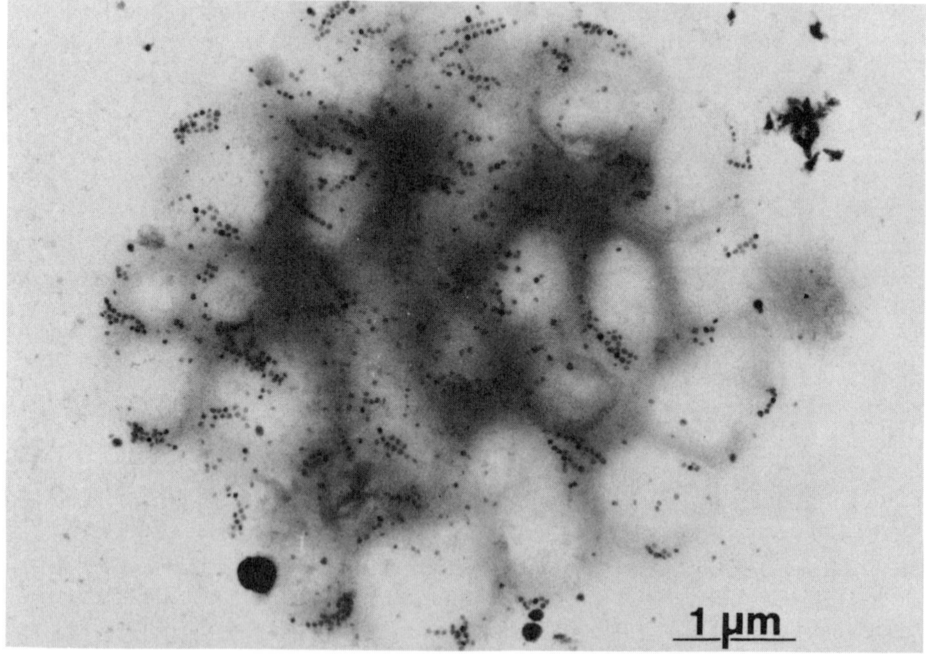

Figure 9. TEM/BF image of a magnetotactic, multicellular prokaryote ("mulberry") showing individual cells and chains of particle (Ref. 21).

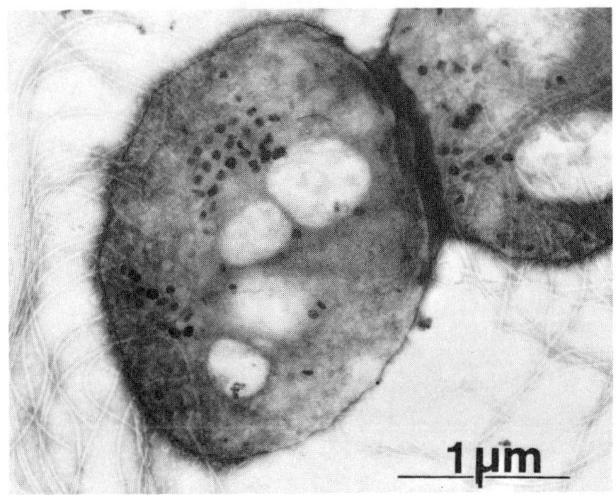

Figure 10. Electron micrograph of one of 20 constituent cells of the magnetotactic, multicellular prokaryote ("mulberry"). Each cell is flagellated and contains magnetosomes. The cell has been negatively stained (Ref. 19).

electron microscopical evidence that the magnetosome chains are oriented parallel to each other in the intact organism. The organism is motile and magnetotactic, but if it is disrupted, e.g., by osmotic shock, the individual cells are not motile, but are permanently magnetic.

As in the magnetotactic rods, elemental mapping shows that the electron-dense inclusions in this organism are iron sulfides (Figure 11).[21,22] However, electron diffraction data reveal that in addition to greigite, the cells contain nonmagnetic pyrite, FeS_2.[21] Since the morphologies of the particles in this organism are not uniform as in the magnetotactic rods, it is not possible to distinguish the particles of greigite and pyrite on the basis of simple examination of the transmission electron micrographs. Thus it is not known how the greigite and pyrite particles are distributed in the cells. However, the electron diffraction data suggest that pyrite is the predominant mineral. The role of non-magnetic pyrite in the cells is not known, but it clearly plays no role in magnetotaxis. Williams[23] has proposed that pyrite is involved in maintenance of iron and sulfide homeostasis in the organisms. Moreover, it has recently been found that copper

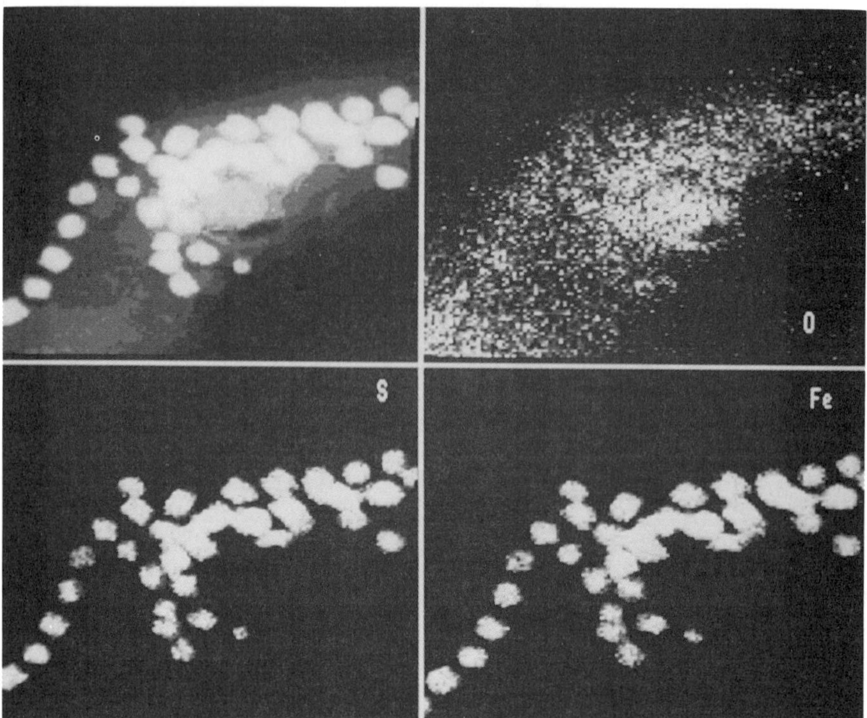

Figure 11. Elemental density maps for the particles in one of the constituent cells of the magnetotactic, multicellular prokaryote: upper left, transmission electron micrograph; upper right, O; bottom left, S; bottom right, Fe. Particles consist of greigite, Fe_3S_4, and pyrite, FeS_2 (Ref. 21). *(See also **Color Plate 5**.)*

can be incorporated into the greigite and/or pyrite particles,[24] suggesting a possible role in detoxification. Nevertheless, it is remarkable that these organisms are separately mineralizing two iron-sulfide minerals, probably in each of its constituent cells.

4.0 CONCLUSION

The rationale for biomimicking as a strategy for the design of novel materials and structures is that the materials and structures found in organisms have been refined and optimized in the course of evolution to have certain properties and to serve specific functions. Even if the functions are not completely understood, the properties can be determined and correlated with the structures. Thus strategies for the production of new materials and structures with similar properties can be determined. Much of the work in this area has focused on the mechanical properties of materials and structures in higher organisms, such as shell, bone, connective tissue, chitin, etc. The magnetotactic bacteria afford an example of optimization of magnetic properties in biomineralized magnetic particles by maintenance of control over particle size and morphology, and creation of a hierarchical structure by relative particle placement in the cell. It is remarkable that the same general scheme is utilized for two different magnetic minerals in different species of bacteria. It has recently been determined by phylogenetic analysis that magnetite-producing magnetotactic bacteria and the iron-sulfide-producing mulberry belong to distinct lineages of the proteobacteria in the domain bacteria.[25] This suggests that the biochemical basis of the biomineralization process could be different for the two mineral particles. In any case, elucidation of the details of bacterial biomineralization processes could yield clues to the production of ordered arrays of inorganic magnetic materials on a nanometer scale.

5.0 ACKNOWLEDGEMENTS

We gratefully acknowledge the participation of A. Garratt-Reed, B.R. Heywood, S. Mann, and N. Sparks in various aspects of this research. We also thank the Office of Naval Research for support.

6.0 REFERENCES

1. R. P. Blakemore, "Magnetotactic Bacteria," *Annu. Rev. Microbiol.*, **36**, 217-238 (1982).

2. R. B. Frankel, "Magnetic Guidance of Organisms," *Annu. Rev. Biophys. Bioeng.*, **13**, 85-103 (1984).

3. R. B. Frankel and R. P. Blakemore, "Magnetite and Magnetotaxis in Microorganisms," *Bioelectromagnetics*, **10**, 223-237 (1989).

4. D. A. Bazylinski, R. B. Frankel, and H. W. Jannasch, "Anaerobic Magnetite Produc-tion by a Marine Magnetotactic Microorganism," *Nature,* **333**, 518-519 (1988).

5. R. P. Blakemore, N. A. Blakemore, D. A. Bazylinski, and T. T. Moench, "The Magnetotactic Bacteria," in *Bergey's Manual of Systematic Bacteriology,* Vol. 3, M. P. Bryant, N. Pfennig, and H. T. Staley (eds.) (Williams and Wilkins, Baltimore, 1989) pp. 15-25.

6. S. Mann and R. B. Frankel, "Magnetite Biomineralization in Unicellular Microorganisms," in *Biomineralisation: Chemical and Biochemical Perspectives*, S. Mann, J. Webb, and R. J. P. Williams (eds.) (VCH, Weinheim, 1989) pp. 388-426.

7. F. Meldrum, S. Mann, B. R. Heywood, R. B. Frankel, and D. A. Bazylinski, "Electron Microscopy of Magnetosomes in Two Cultured Vibrioid Magnetotactic Bacteria," *Proc. Roy. Soc. B,* **251** 237-242 (1993).

8. N. H. C. Sparks, S. Mann, D. A. Bazylinski, D. R. Lovley, H. W. Jannasch, and R. B. Frankel, "Structure and Morphology of Anaerobically Produced Magnetite by a Marine Magnetotactic Bacterium and a Dissimilatory Iron-Reducing Bacterium," *Earth Planet. Sci. Lett.,* **98**, 14-22 (1990).

9. H. A. Lowenstam, "Minerals Formed by Organisms," *Science,* **211**, 1126-1131 (1981).

10. S. Mann, "On the Nature of Boundary Organised Biomineralization," *J. Inorg. Biochem.*, **28**, 363-370 (1986).

11. Y. A. Gorby, T. J. Beveridge, and R. P. Blakemore, "Characterization of the Bacterial Magnetosome Membrane," *J. Bacteriol.*, **170**, 834-841 (1988).

12. R. F. Butler and S. K. Banerjee, "Theoretical Single-Domain Grain Size Range in Magnetite and Titanomagnetite," *J. Geophys. Res.*, **80**, 4049-4058 (1975).

13. R. P. Blakemore, K. A. Short, D. A. Bazylinski, C. Rosenblatt, and R. B. Frankel, "Microaerobic Conditions are Required for Magnetite Formation within *Aquaspirillum magnetotacticum,*" *Geomicrobiol. J.*, **4**, 53-71 (1985).

14. D. R. Lovley, J. F. Stoltz, G. L. Nord, Jr., and E. J. P. Phillips, "Anaerobic Production of Magnetite by a Dissimilatory Iron-Reducing Microorganism," *Nature,* **330**, 252-254 (1987).

15. D. A. Bazylinski, R. B. Frankel, A. J. Garratt-Reed, and S. Mann, "Biomineralization of Iron Sulfides by Magnetotactic Bacteria from Sulfidic Environments," in *Iron Biominerals*, R. B. Frankel and R. P. Blakemore (eds.) (Plenum, New York, 1991) pp. 239-255.

16. B. R. Heywood, D. A. Bazylinski, A. Garratt-Reed, S. Mann, and R. B. Frankel, "Controlled Biosynthesis of Greigite (Fe_3S_4) in Magnetotactic Bacteria," *Naturwissenschaften*, **77**, 536-538 (1990).

17. B. R. Heywood, S. Mann, and R. B. Frankel, "Structure, Morphology, and Growth of Biogenic Greigite (Fe_3S_4)," in *Materials Syntheses Based on Biological Processes*, M. Alper, P. Calvert, R. Frankel, P. Rieke, and D. Tirrell (eds.) Proc. of MRS, Vol. 218 (Materials Research Society, Pittsburgh, 1991) pp. 93-108.

18. D. A. Bazylinski, "Bacterial Production of Iron Sulfides," in *Materials Syntheses Based on Biological Processes*, M. Alper, P. Calvert, R. Frankel, P. Rieke, and D. Tirrell (eds.) Proc. of MRS, Vol. 218 (Materials Research Society, Pittsburgh, 1991) pp. 81-91.

19. F. Rogers, R. P. Blakemore, N. A. Blakemore, R. B. Frankel, D. A. Bazylinski, D. Maratea, and C. Rogers, "Intercellular Structure in a Many-celled Magnetotactic Procaryote," *Arch. Microbiol.*, **154**, 18-22 (1990).

20. M. Farina, H. Lins de Barros, D. Motta de Esquivel, and J. Danon, "Ultrastructure of a Magnetotactic Microorganism," *Biol. Cell*, **48**, 85-88 (1983).

21. S. Mann, N. H. C. Sparks, R. B. Frankel, D. A. Bazylinski, and H. W. Jannasch, "Biomineralization of Ferrimagnetic Greigite (Fe_3S_4) and Iron Pyrite (FeS_2) in a Magnetotactic Bacterium," *Nature,* **343**, 258-260 (1990).

22. M. Farina, D. Motta de Esquivel, and H. G. P. Lins de Barros, "Magnetic Iron-Sulphur Crystals from a Magnetotactic Microorganism," *Nature,* **343**, 256-258 (1990).

23. R. J. P. Williams, "Biominerals and Homeostasis," in *Iron Biominerals*, R. B. Frankel and R. P. Blakemore (eds.) (Plenum, New York, 1991) pp. 7-19.

24. D. A. Bazylinski, A. J. Garratt-Reed, A. Abedi, and R. B. Frankel, "Copper Association with Iron Magnetosomes in a Magnetotactic Bacterium," *Arch. Microbiol.*, **60** 35-42 (1993).

25. E. F. DeLong, R. B. Frankel, and D. A. Bazylinski, "Multiple Evolutionary Origins of Magnetotaxis in Bacteria," *Science,* **259** 803-806 (1993).

SIZE-QUANTIZED PARTICLES AT ARTIFICIAL MEMBRANE INTERFACES

Janos H. Fendler

Department of Chemistry, Syracuse University
Syracuse, New York 13244-4100, USA

CdS and ZnS semiconductor particulate films, prepared at negatively charged monolayer interfaces and transferred to solid supports, have been characterized by reflectivity, absorption spectrophotometry, transmission electron microscopy, scanning tunneling microscopy, and electrical measurements. Plots of absorbances at a given wavelength against thickness were linear for CdS and ZnS particulate films. Direct bandgaps for 63-, 125-, 163-, 204-, 263-, and 298-Å-thick, CdS particulate films were evaluated to be 2.54, 2.48, 2.46, 2.44, 2.43, and 2.42 eV, respectively. Similarly, a direct band-gap of 3.75 eV was assessed for the 458-Å-thick, ZnS particulate film. Transmission electron micrographs of CdS films revealed the presence of CdS particles in a narrow size distribution with average diameters of 47 Å. The presence of 20- to 30-Å-thick, 40- to 50- Å-diameter CdS and 10- to 25-Å-thick, 30- to 40-Å-diameter ZnS particles in CdS and ZnS films were discerned by scanning tunneling microscopy. CdS films had dark resistivities of (3-6)10^7 Ωcm, which decreased upon illumination; they also developed photovoltages upon illumination. The action spectrum of the photoconductivity corresponded to the absorption spectrum of the CdS particulate film, indicating its origin to be the conduction band electrons and valence band holes produced in band-gap excitation.

1.0 INTRODUCTION

Systematic investigation of colloidal semiconductor particles has revealed important size- and dimensionality-dependent properties.[1-4] In particular, quantization of the energy levels became observable in particles which were smaller than the exciton diameter of the bulk semiconductor. This electron-hole confinement in spherical particles, the three-dimensional quantum size effect, resulted in the formation of "quantum dots" or "quantum crystallites" or "zero-dimensional excitons." Interest in three-dimensional size quantization has been prompted by altered mechanical, chemical, electrical, and electrooptical properties which may be potentially exploited in a variety of applications.[5]

Nanosized particles have been prepared by mechanical reduction of larger structures,[6] by direct chemical synthesis,[7] and by colloid chemical techniques using controlled mixing of the precursors under well-adjusted conditions.[8-23] Maintaining monodispersed particles in controlled sizes is experimentally demanding. Nonaqueous solvents, low temperature, stabilizers, and chemical treatments (capping) have been used for the stabilization of ultrasmall semiconductor particles. In our laboratory, we have taken advantage of organized surfactant aggregates as matrices for the stabilization of colloidal semiconductor particles. In particular, CdS and related semiconductor particles have been *in situ* generated in reversed micelles,[13] surfactant and polymerized surfactant vesicles,[14,24-29] bilayer lipid membranes (BLMs),[21,30,31] and between the head-groups of Langmuir-Blodgett (LB) films.[32] The general methodology involved the slow exposure of the appropriate metal cation precursor, electrostatically attracted to the oppositely charged surfactant aggregate surface, to H_2S. Particularly sensitive control was provided by the BLM since it allowed the spatial separation of the metal-ion precursor and H_2S by the membrane.[21] Coating only one side of the BLM by metal ions (the *cis* side) and infusing H_2S from the aqueous solution bathing the opposite side (the *trans* side) of the bilayer led to the gradual appearance of ultrasmall semiconductor particles on the *cis* side of the membrane. The semiconductor particles rapidly moved around in the matrix of the membrane and grew in size, forming islands which merged with themselves and with a second generation of particles. Growth ultimately led to a continuous, porous semiconductor particulate film which grew in thickness in the direction perpendicular to the BLM. The growth of the semiconductor particulate film could be controlled by the amount and the rate of H_2S infusion.[21,30,32] Size-quantized semiconductor particles have also been incorporated between the hydrophilic layers of LB films.[33]

The generation of semiconductor particles from their metal-ion precursors electrostatically attracted to negatively charged monolayers, their transfer to solid substrates, and their characterization is the subject of the present chapter.

2.0 EXPERIMENTAL APPROACH

Bovine brain phosphatidylserine, PS (Avanti Polar Lipids, Inc.), arachidic acid, AA (Sigma Chemical Co.), dioctadecyldimethylammonium bromide, DODAB (Eastman), cadmium chloride, zinc chloride, copper sulfate (Fisher), high-purity dry N_2 (Union Carbide), H_2S (Matheson), and spectroscopic grade chloroform (Aldrich) were used as received. Preparation and purification of polymerizable surfactants n-hexadecyl-11-(vinylbenzamide) undecyl hydrogen phosphate, **1**; bis(2n-hexadecanoyloxyethyl) methyl(p-vinylbenzyl)ammonium chloride, **2**; and poly(styrenephosphonate diester), PSP, have been described in a previous publication.[34] Water was purified by a Millipore Milli-Q filter system provided with a 0.22 μm Millistack filter at the outlet.

Semiconductor particles were generated *in situ* at the monolayer headgroup-aqueous subphase interface. An aqueous metal ion solution (1.0 x 10^{-3} M $CdCl_2$, or $ZnCl_2$, or $CuSO_4$) constituted the subphase. Either a commercial Lauda Model P Langmuir film balance or a simple circular trough was used for monolayer formation and the subsequent generation of semiconductor particles. The Lauda film balance was enclosed in a plexiglass hood and placed on a Micro-g optical isolation table. The water surface was cleaned several times by sweeping with a teflon barrier prior to monolayer formation. The subphase was deemed to be clean when the surface pressure increase was less than 0.2 dyne/cm upon compression to 1/20 of the original area and when this surface pressure increase remained the same subsequent to aging for several hours. An appropriate amount of chloroform solution of the chosen surfactant (8 x 10^{17} molecules per mL) was carefully injected onto the clean, thermostated (25.0°), aqueous surface. The surfactants were compressed at a rate of (2-5)10^{-3} Å^2 per molecule per second. After 5 to 15 min of incubation at the desired surface pressure (25 mN/m for AA and 50 mN/m for PS, DODAB, **1** and **2**), 200 to 250 μL H_2S was slowly injected into the nitrogen-filled plexiglass hood covering the film balance. Semiconductor particle formation at the monolayer surface was visually observed and the particulate films formed were transferred onto solid substrates by horizontal lifting for subsequent characterization.[35]

The schematics of the circular, 4.0 cm deep, pyrex trough (7.0 cm^2 surface area) used for *in situ* semiconductor particle generation are shown in Figure 1. The trough was placed on a clean (chromic acid, copious amounts of water) flat glass plate and covered

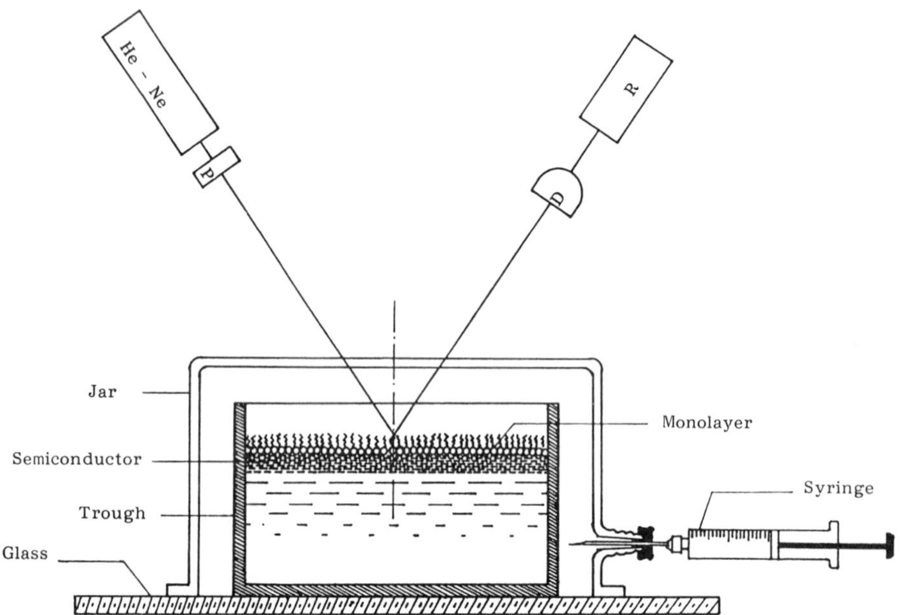

Figure 1. Schematics of the experimental arrangements used for the generation of semiconductor particles at the negatively charged, surfactant headgroup-aqueous (1.0 x 10^{-3} M MCl_2) subphase interface and that used for the *in situ* monitoring of reflectivities. P= polarizer, D= detector, R= chart recorder.

by a circular (7.5 cm high, 14.5 cm in diameter) glass jar whose flat frosted bottom provided a gas-tight contact. The water surface was cleaned by sweeping it through a water aspirator. An appropriate amount of the spreading solution (1.5 x 10^{-3} M surfactant in $CHCl_3$) was carefully injected onto the clean subphase to give a coverage of 20 $Å^2$ per molecule of AA and 40 $Å^2$ per molecule of PS, DODAB, **1** and **2**. Subsequent to 20 minutes of incubation, a Hamilton syringe containing 200 to 500 μL H_2S was introduced into the atmosphere covering the monolayer via a rubber septum (Figure 1). The barrel of the syringe was kept at the same position. This ensured the extremely slow (several hours) diffusion of H_2S into the chamber and, hence, into the monolayer-water interface. The formation of semiconductor particles was monitored by reflectivity measurements (*vide infra*). The semiconductor particulate films formed were transferred onto solid substrates by horizontal lifting for subsequent characterization.[35]

The experimental arrangements for *in situ* optical reflectivity measurements are illustrated in Figure 1. A He-Ne laser (Hughes, 13 mW) was tightly mounted on a precision rotating optical stage. The rotator with the rotating axis in the horizontal direction was supported by an optical post. Care was taken to provide good vibration isolation during measurements. Parallel-polarized laser light (6328 Å) was directed to the monolayer surface. For incident angle measurements, the angle of incident light, θ, was changed by sliding the post in a horizontal direction parallel to the surface without affecting the optical geometry. The intensity of the incident light, corresponding to 100% reflectivity, was measured with $2\theta = 180°$. The intensity of the reflected light was measured by means of a Spectra Physics Model 404 silicon photocell power meter which was connected to a chart recorder.

The monolayer-supported semiconductor particulate film was transferred to solid substrates by horizontal lifting through the surface layers. Well-cleaned (chromic acid, dust-free water), 1.0 cm x 4.5 cm x 0.1 cm, spectroscopic-grade quartz plates; right-angle glass prisms; cellulose-nitrate-coated copper grids; highly oriented, pyrolytic graphite (HOPG, Union Carbide, freshly cleaved); and glass slides or teflon sheets (etched cuflon, CF-A-01-5-5, Polyflon Corporation) were used as substrates for absorption spectrophotometry, reflectivity, transmission electron microscopy (TEM), scanning tunneling microscopy (STM), and electrical measurements, respectively. Absorption spectra were taken either on a Hewlett-Packard 8450 A diode-array spectrophotometer or on a PV 8800 Philips spectrophotometer.

TEM images were taken with a JEOL JEM-2000 EX 200 keV instrument. Samples were obtained on specially prepared substrates. Several pieces of 200-mesh copper grids were placed on a glass slide (1.0 cm x 4.5 cm x 0.1 cm) and immersed horizontally under the surface of purified water. A drop of cellulose nitrate (1.0% in amyl acetate, Ernest F. Fullam, Inc.) was allowed to spread evenly on the water surface. Subsequent to the formation of a thin, cellulose-nitrate film, the glass slide carrying the copper grids was horizontally lifted through the cellulose-nitrate-coated water surface. Drying in a vacuum desiccator for 1 d produced the copper grid substrates used in TEM measurements.

STM images were acquired by means of an Angstrom Technology (Mesa, Arizona) TAK 2.0 instrument operated in the constant current mode. A Pt-Ir wire was used for the tunneling tip. Images were scanned with five lines per second and 0.5 to 1.0 volts tip bias (images were plotted on a CP 200U Mitsubishi color videocopy processor). Subsequent to transferring to graphite, the films were dried for 24 h and surfactant monolayers were removed by gentle rinsing with $CHCl_3$, ethanol, and dust-free water. Eight separately prepared samples of CdS and ZnS were investigated. Images were taken in each sample in 50 to 100 different areas.

All electrical measurements were performed in a Faraday cage. Glass-supported semiconductor particulate films had two parallel gold or indium electrodes deposited in a 10^{-6} torr evaporator to an 800 Å thickness prior or subsequent to the film transfer. Teflon-supported semiconductor particulate films had a copper (etched from cuflon prior to the film transfer) and a gold (or indium) electrode (vacuum deposited subsequent to the film transfer). Steady-state photovoltage and photoconductivity measurements were performed by connecting the electrodes to a Keithley Model 602 electrometer with an input impedance of 10^{14} Ω. Solid-supported particulate films were dried in a vacuum chamber (10^{-3} torr) for 3 d prior to the electrical measurements. Samples were irradiated at a 45° angle by a 150 W Xenon lamp via an ultraviolet sensitive optical fiber. Output from the electrometer was connected to a chart recorder. When needed, monochromatic light was provided by placing a Spex 1681, 0.22 m monochromator into the excitation light path. The energy per area (effected by the monochromatic light at a given wavelength) was determined by a Scientec 3652 energy/power meter.

Time-dependent photovoltage was investigated (also at 45° incident angle of irradiation) by using 10 nsec, 343 nm laser pulses (Lambda Physik EMG 101 MS excimer laser pumping a Lambda Physik 2002 dye laser; p-terphenyl). The two electrodes were directly connected to the input of a Tektronix 4662 storage oscilloscope in order to improve the response time. The oscilloscope scanning was synchronized with the laser pulse via a trigger and digital delay generator (Model 111 AR) with nanosecond accuracy. Photovoltage signals were transferred to a Zenith Data System computer (Z-100) via a programmable digitizer (Sony-Tektronix 390 AD). The instrument response time was 10^{-8} sec. The energy of the laser pulse, varied by neutral density filters, was determined by means of a Scientec 3652 power/energy meter.

Temperature-dependent resistivities were determined by connecting the sample covered substrate to a brass base. The brass was then heated by an AC power supply via a heating wire. The temperature was measured by a point contact thermocouple.

3.0 RESULTS AND DISCUSSION

3.1 SURFACE AREA-SURFACE PRESSURE ISOTHERMS AND SEMICONDUCTOR PARTICLE GENERATION

Surface pressure-surface area isotherms of monolayers prepared from AA, PS, DODAB, **1** and **2**, showed the expected behavior (Figure 2). At low surface pressure, molecules lying flat on the subphase occupied large areas. With increasing surface pressure, they began to be squeezed together and to orient their hydrophobic tails away from the surface. Transition to a compressed (solid) state manifested in a rapid rise of the surface pressure with relatively little change in the surface area. This solid state prevailed until the collapse of the monolayer. Depending on the subphase (1.0×10^{-3} M $CdCl_2$, or $ZnCl_2$, or $CuSO_4$), collapsed pressures of monolayers prepared from AA, PS, **1**, DODAB, and **2** were found to be 30 to 40 mN/m, 60 to 70 mN/m, 60 to 70 mN/m, 48 to 58 mN/m, and 44 to 54 mN/m, respectively. These values were in the expected range.[36] Semiconductor particle formation, initiated by the introduction of H_2S into the hood covering the trough (see Figure 1 and Experimental section), was observable in compressed and incubated monolayers prepared from negatively charged AA, PS, and **1** monolayers. In contrast, semiconductor formation could not be observed in monolayers prepared from positively charged DODAB and **2**. Infusion of H_2S over the aqueous subphase in the absence of monolayers led to the formation of large quantities of yellowish (CdS), whitish (ZnS), and dark-brownish (CuS-Cu_2S) particles which precipitated in the aqueous solution and settled at the bottom of the trough. A typical pressure-area isotherm is shown in Figure 2. Significantly, no precipitation occurred on using negatively charged monolayers. Nucleation at the surfactant headgroups was followed by two dimensional particle growth to cover the entire monolayer. Subsequently, the particulate semiconductor film grew in thickness perpendicular to the plane of the monolayer. All of the incipient particles were supported by

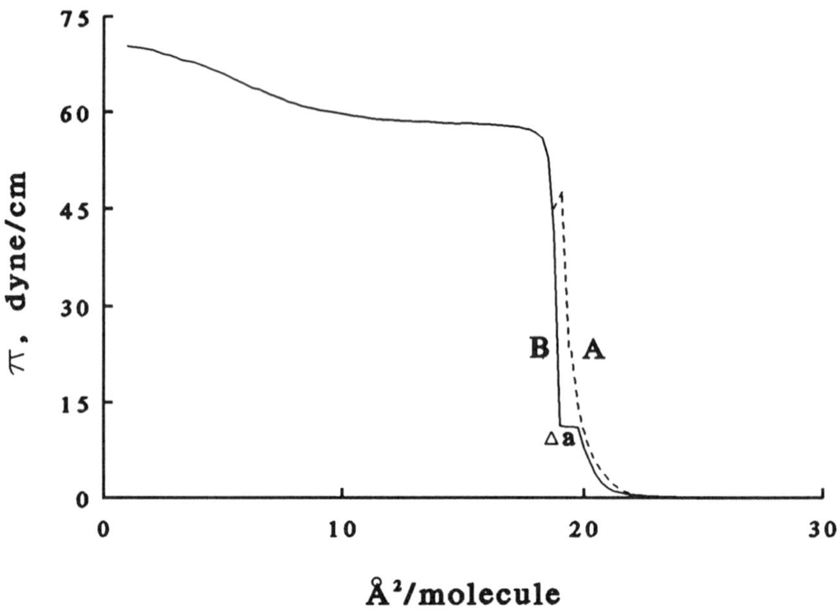

Figure 2. Surface pressure-surface area isotherm of arachidic acid on 2.5 x 10^{-4} M $CdCl_2$ prior (A) and five minutes subsequent (B) to the introduction of H_2S over the monolayer. The area loss, Δa, corresponds to the transfer of the monolayer to a substrate at $\pi = 11.5\ mNm^{-1}$ = constant.

the monolayer (i.e., there were no colloidal semiconductor particles in the aqueous subphase). Strong electrostatic and particle-particle interactions are, presumably, responsible for maintaining the particulate semiconductor film at the monolayer interface.

3.2 OPTICAL THICKNESS AND MORPHOLOGY OF SEMICONDUCTOR PARTICULATE FILMS ON SOLID SUPPORTS

The uniform growth of semiconductor particles at the monolayer interface, ensured by the controlled and slow infusion of H_2S, was monitored by reflectivity measurements at $\Theta = 10°$ incident angle.[35] With continuous slow H_2S infusion, the intensity of the light reflected from the monolayer-supported CdS film increased to a plateau value which corresponded to a limiting thickness of 225 to 300 Å. A different behavior was found in the generation of ZnS particles. Sequential color changes, white→yellow→gold→orange→pink→violet→blue→green→yellow and gold, were accompanied by interference maxima and minima.[35] The effective refraction indices, n_s, and volume fractions, F, of arachidate-monolayer-supported semiconductor particulate films are collected in Table-I.

Table-I. Effective Refractive Indices and Volume Fractions of Semiconductor Particulate Films at Monolayers and on Solid Supports*

Monolayer-Supported	**Solid-Supported**		**Bulk Semi-Conductor**		
	ns	F, %	n's	F', %	nb
CdS	2.14	75.5	2.25	79.5	2.50
ZnS	1.84	55.4	1.88	58.3	2.37

* ns = effective refractive index of the monolayer-supported semiconductor particulate film;
n's = effective refractive index of the substrate-supported semiconductor particulate film;
F = volume fraction of the monolayer-supported semiconductor particulate film;
F' = volume fraction of the substrate-supported semiconductor particulate film.

The monolayer-supported semiconductor particulate films were quite fragile. Vertical insertion of a stainless steel needle or a solid substrate caused a crack which grew to several centimeters in length. Traditional vertical Langmuir-Blodgett lifting could not be employed, therefore, for effecting transfer to solid substrates. In contrast, horizontal lifting of substrates resulted in the intact and quantitative deposition of the monolayer-supported, semiconductor particulate films. Any one of the 200- to 300-Å-thick CdS and 200- to 2000-Å-thick ZnS films appeared to be uniform and remained stable even on 200-mesh copper grids. Conversely, ZnS films thicker than 2000 Å cracked on drying.

Incident-angle-dependent reflectivities of differently colored ZnS particulate films on glass prisms (to obviate substrate interference), prepared by exposing zinc-arachidate monolayers to H_2S for increasingly longer amounts of time, are shown in Figure 3. The intersection of the curves in Figure 3 defined the Brewster angle, $\Theta_B = 62.0°$, which allowed the assessment of the effective refractive index, n'_s, from

$$n_s = \tan \Theta_B \tag{1}$$

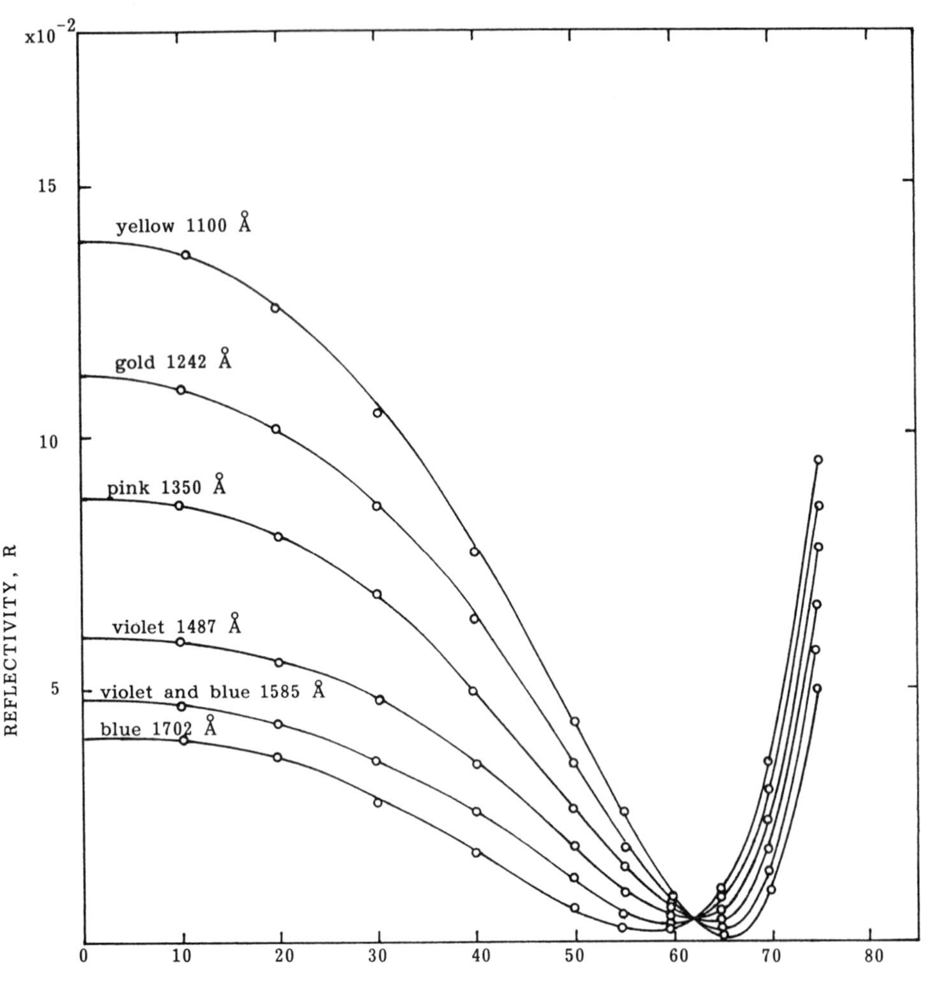

Figure 3. Incident-angle-dependent reflectivities of ZnS particulate films on glass prisms. The different curves were taken at different times subsequent to the beginning of H_2S infusion onto the zinc arachidate monolayers. The colors and calculated thicknesses are indicated on the curves. The intersection of the curves, 62.0°, defines the Brewster angle, Θ_B.

to be 1.88. This, in turn, permitted the numerical evaluation of the optical thicknesses of the semiconductor particulate films on the glass prism.

3.3 ABSORPTION SPECTROPHOTOMETRY

Absorption spectra of semiconductor particulate films on quartz substrates were found to be highly sensitive to the method used in their preparation. Specifically, well-compressed cadmium-arachidate-, zinc-arachidate-, or metal-ion-coated PSP monolayers; an appropriate subphase; and, most importantly, slow infusion of H_2S for a limited time were required for shifting the absorption threshold to higher energies. Typical absorption spectra of 8-, 35-, 218-, 298-, and 328-Å-thick CdS particulate films, *in situ* generated at cadmium-arachidate monolayer/aqueous 1.0 x 10^{-3} M $CdSO_4$ inter-faces, are shown in Figure 4.

The relationship between absorption spectra of CdS particles and their sizes has been studied extensively.[1,9] In particular, Henglein and co-workers prepared CdS particles with different mean diameters and obtained a curvilinear relationship between their absorption thresholds (λ_t) and their independently determined sizes.[37-39] The thinnest particulate CdS films (8 Å thick), prepared from Cd-arachidate monolayers, had an absorption threshold (λ_t) of 450 nm (Figure 4). This value led to an assessment of 30±5 Å for the mean diameter of CdS particles on using Henglein's λ_t versus particle size curve,[37] in good agreement with those obtained in TEM measurements.

The absorption spectra of semiconductor particulate films thicker than 30 to 40 Å showed long tails near their absorption edges (Figure 4). These long tails could be due to defect states, indirect transitions, and particle size distribution. Theoretical calculations for semiconductor crystallites established that the square of absorption coefficients times energy, $(\sigma h\omega)^2$, increases gradually above the band-gap (E_0 transition) with increasing energy. This increase can be approximated by:

$$(\sigma h\omega)^2 = (h\omega - E_g)C \tag{2}$$

where $h\omega$ is the photon energy, E_g is the direct band-gap, and c is a constant. Reflectivity measurements provided values for the thickness (d'_s) of the semiconductor particulate films, which permitted the calculation of $\sigma(\sigma = A/d'_s$, where A is the measured

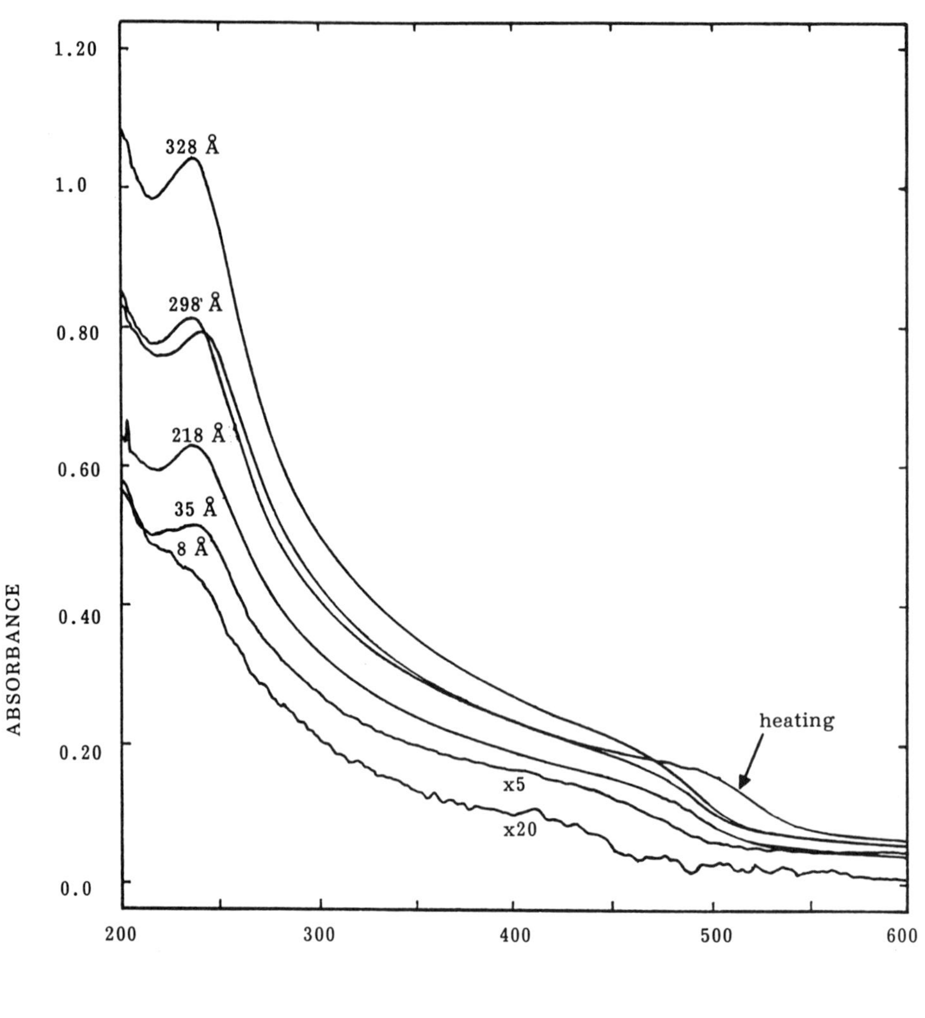

Figure 4. Absorption spectra of 8-, 35-, 218-, 298-, and 328-Å-thick (indicated on the curves) CdS particulate films on quartz supports. The arrow indicates the absorption spectrum of a 328-Å-thick, quart-supported CdS particulate film after it was heated at 400°C for 20 min. Particulate films were prepared by the infusion of H_2S onto a cadmium-arachidate monolayer.

absorbance) and, upon substitution into Eq. 2, led to the plot of $(\sigma h\omega)^2$ against $h\omega$. Ignoring the weak absorption tail, as is customary,[39] resulted in the precise determination of direct band-gaps by using extrapolated values. The proposed morphological evolution of semiconductor particulate films is illustrated in Figure 5.

A TEM of a CdS semiconductor particulate film, *in situ* generated at a cadmium ion-PSP monolayer interface by 3 min of exposure to H_2S, is shown in Figure 6. The shorter the H_2S exposure time is, the smaller the size is of the semiconductor particle that is formed. Those formed at cadmium-arachidate interfaces after 3 min of H_2S exposure are not visible, in fact, by the naked eye and correspond to CdS particles of 30±5 Å diameters (Figure 6). Nascent individual semiconductor particles are well separated at their earliest stages of growth. Longer exposures to H_2S result in the aggregation ("clumping") of the smaller particles, in the appearance of larger ones, and, ultimately, in chain and disk formation. Individual small crystallites in the larger particles are resolved in some of the electromicrograms. A similar situation has been encountered in the formation of ZnS semiconductor particulate films.

3.4 SCANNING TUNNELING MICROSCOPY STUDIES OF THIN FILMS OF PARTICLES

Evolution of ZnS particulate films from their earliest stages of growth can be convincingly seen in scanning tunneling microscopy (STM) images. The initially formed (typically 2 to 6 min of H_2S exposure), well-separated, 20- to 30-Å diameter and 5- to 6-Å-high nanoclusters (Figure 7) grew quickly in height to longer and more densely packed, conical particles. Longer exposure (>20 min) to H_2S resulted in increased lateral growth of the nanocrystallites and their clumping into interconnected arrays. Quite often, individual nanocrystallites of chained or disk-shaped semiconductor particulate assemblies are visible in the STM images. Large CdS clusters were also demonstrated to be composed of much smaller (15- to 40-Å-diameter) nanoparticles by STM.[40] Prolonged exposure (>30 min) ultimately lead to the formation of a "first layer" of particulate semiconductor film. Inspection of STM images at this stage revealed the presence of 30- to 40-Å-thick, 30- to 80-Å-diameter, disk-shaped semiconductor particles.

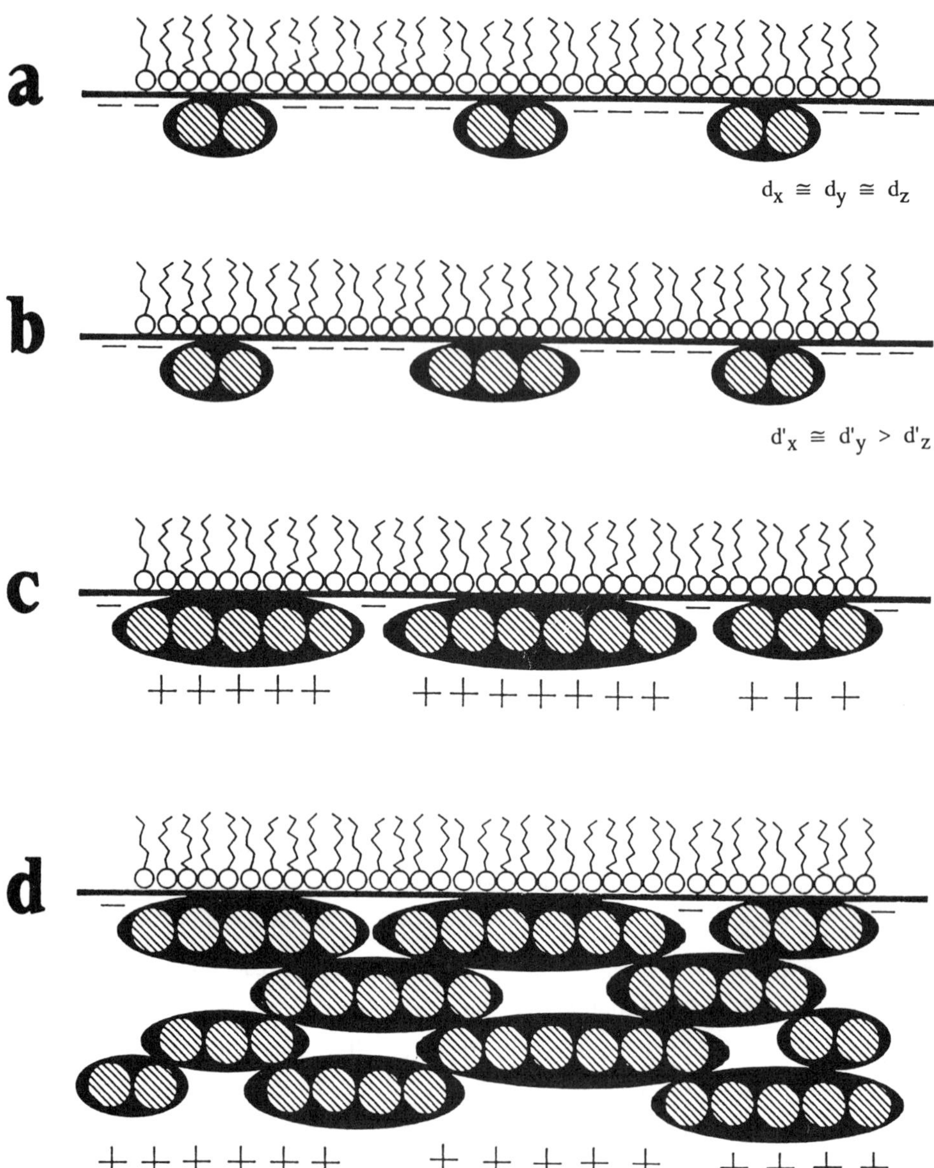

Figure 5. Proposed schematics for the initial (a) and subsequent (b,c, and d) growth of a monolayer-supported, porous, semiconductor particulate film. The d_x and d_y dimensions are in the plane and the dz dimension is normal to the plane; they refer to the earliest observable particles. d'x, d'y, and d'z are dimensions in the plane and are normal to the plane; they refer to particles observed at later stages of their growth.

Figure 6. A TEM image of CdS particles, *in situ* formed at a PSP monolayer by exposure to H_2S for 3 min.

3.5 ELECTRICAL AND PHOTOELECTRICAL PROPERTIES OF SEMICONDUCTOR PARTICULATE FILMS ON SOLID SUPPORTS

Electrical and photoelectrical measurements were carried out on CdS particulate films deposited on glass substrates or teflon sheets. The resistivity (ρ) of a semiconductor particulate film, measured between two parallel copper electrodes, is given by

$$\rho = R\frac{Ld'_s}{a} \tag{3}$$

where R is the measured resistivity in Ω, L is the length of the copper electrodes, a is the distance between them, and d'_s is the thickness of the semiconductor particulate film. Resistivities of 200-to 300-Å-thick, CdS particulate films were determined to be $(3\text{-}6)10^7$ Ω cm. The range represents measurements of 10 samples of different thicknesses and is due in part to the presence of different amounts of water in the films. The ρ values determined for CdS particulate films are some six orders of magnitude higher than those observed for materials having intrinsic conductivity.

The dark resistance of CdS particulate films was found to decrease exponentially increasing temperature. Illumination decreased the resistivity (i.e., increased the conductivity) of CdS particulate films by some two orders of magnitude and matched the absorption spectrum of the corresponding CdS particulate film nicely. Photoconduc-

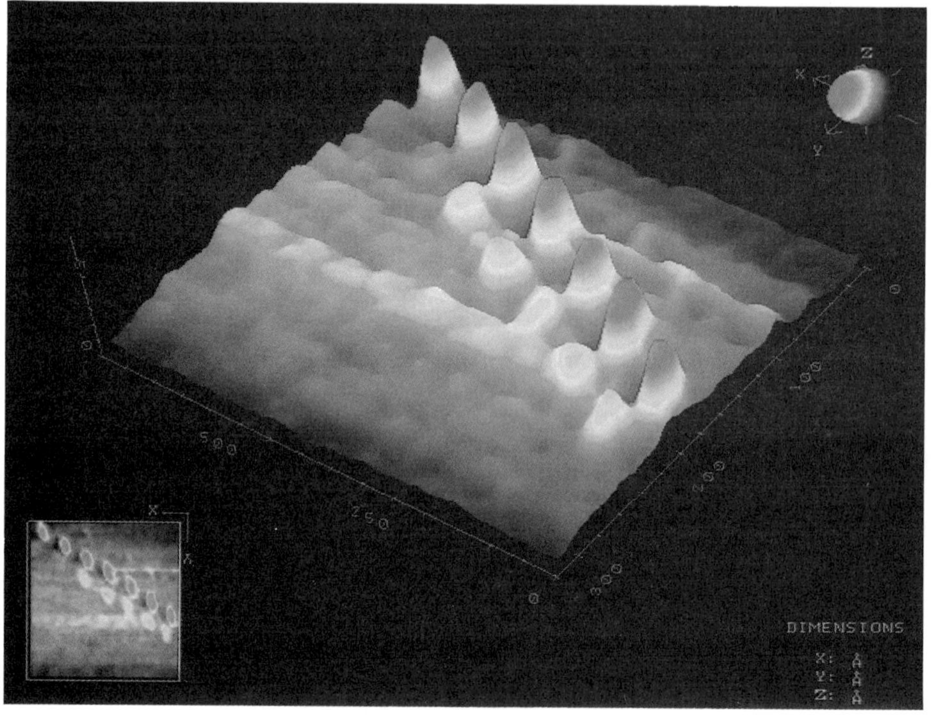

Figure 7. A three-dimensional STM image of ZnS particles on HOPG, *in situ* generated on a zinc-arachidate monolayer by 5 min exposure to H_2S. A bird's eye view of the same area in two dimensions (x and y axes are shown on the same scale as in the three-dimensional plot) is shown in the insert. *(See also **Color Plate 6**.)*

tivity originates, therefore, in the production of conduction band electrons, $e\text{-}_{CB}$, and valence band holes, $h\text{+}_{VB}$, in band-gap irradiation of CdS:

$$CdS \xrightarrow{hu} e^{-}{}_{CB} + h +_{VB} \tag{4}$$

4.0 CONCLUSION

Semiconductor size quantization has been demonstrated in thin films grown at surfactant monolayer interfaces. The method of preparation permits convenient thickness, morphology, and dimensionality controls. Manipulations of the surfactant organization and the chemistries at the monolayer interface and subphase will provide much better control of semiconductor particulate film formations and permit the colloid chemical generation of periodically and spatially modulated superlattices. Most importantly, substrate-supported, size-quantized semiconductor films can be fully characterized by solid-state methodologies and can be exploited as novel electronic devices.

5.0 ACKNOWLEDGEMENT

Support of this work by a grant from the Department of Energy is gratefully acknowledged.

6.0 REFERENCES

1. A. Henglein, *Topics Curr. Chem.*, **143**, 113 (1988).

2. L. A. Brus, *J. Phys. Chem.*, **90**, 2555 (1986).

3. M. L. Steigerwald and L. E. Brus, *Acc. Chem. Res.*, **23**, 183 (1990).

4. J. H. Fendler, *Chem. Rev.*, **87**, 877 (1987).

5. R. P. Andres, R. S. Averback, W. L. Brown, L. E. Brus, W. A. Goddard, A. Kaldor, S. G. Louie, M. Moskovits, P. S. Percy, S. J. Riley, R. W. Siegel, F. Spaepen, and Y. Wang, *J. Mater. Res.*, **4**, 704 (1989).

6. S. W. Charles and J. Popplewell, in *Ferromagnetic Materials,* Vol. 2, E. P. Wohlfarth (ed.) (North Holland, New York, 1980).

7. M. L. Steigerwald and C. R. Sprinkle, *J. Am. Chem. Soc.,* **109**, 7200 (1987).

8. J. G. Brennan, T. Siegrist, P. J. Caroll, S. M. Stuczynski, L. E. Brus, and M. L. Steigerwald, *J. Am. Chem. Soc.,* **111**, 4141 (1989); D. Meissner, R. Memming, and B. Kastening, *Chem. Phys., Lett.,* **96**, 34 (1983).

9. R. Rosetti, R. Hull, J. M. Gibson, and L. E. Brus, *J. Chem. Phys.*, **82**, 552 (1984).

10. A. W. H. Mau, C. B. Huang, N. Kakuta, A. J. Bard, A. Campion, M. A. Fox, M. J. White, and S. E. Webber, *J. Am. Chem. Soc.*, **106**, 6537 (1984).

11. J. P. Kuczynski, B. M. Milosavljevic, and J. K. Thomas, *J. Phys. Chem.*, **88**, 980 (1984).

12. A. Fojtik, H. Weller, U. Koch, and A. Henglein, *Ber. Bensenges. Phys. Chem.,* **88**, 969 (1984).

13. M. Meyer, C. Wallberg, K. Kurihara, and J. H. Fendler, *J. Am. Chem. Soc., Chem. Commun.,* **90** (1984).

14. H. C. Youn, Y. -M. Tricot, and J. H. Fendler, *J. Phys. Chem.,* **89**, 1236 (1985).

15. H. T. Tien, Z. C. Bi, and A. K. Tripathi, *Photochem. Photobiol.,* **44**, 779 (1986).

16. P. Lianos and J. K. Thomas, *J. Coll. Surf. Sci.*, **117**, 505 (1986).

17. T. Daunhauser, M. O'Neil, K. Johansson, D. Whitten, and G. McLendon, *J. Phys. Chem.*, **90**, 6074 (1986).

18. Y. Wang and W. Mahler, *Opt. Comm.,* **61**, 233 (1987).

19. Y. Wang, A. Suna, W. Mahler, and R. Rasowski, *J. Chem. Phys.,* **87**, 7315 (1987).

20. Y. Wang and N. Herron, *J. Phys. Chem.*, **91**, 257 (1987).

21. X. K. Zhao, S. Baral, R. Rolandi, and J. H. Fendler, *J. Am. Chem. Soc.,* **110**, 1012 (1988).

22. C. Petit and M. P. Pileni, *J. Phys. Chem.*, **92**, 2282 (1988).

23. M. L. Steigerwald, A. P. Alivisatos, J. M. Gibson, T. D. Harris, R. Kortan, A. J. Muller, A. M. Tayer, T. M. Duncan, D. C. Douglass, and L. E. Brus, *J. Am. Chem. Soc.*, **110**, 3046 (1988).

24. Y. -M Tricot and J. H. Fendler, *J. Am. Chem. Soc.*, **106**, 2475 (1984).

25. Y. -M. Tricot and J. H. Fendler, *J. Am. Chem. Soc.*, **106**, 7359 (1984).

26. R. Rafaeloff, Y. -M. Tricot, F. Nome, and J. H. Fendler, *J. Phys. Chem.*, **89**, 533 (1985).

27. R. Rafaeloff, Y. -M. Tricot, F. Nome, P. Tundo, and J. H. Fendler, *J. Phys., Chem.* **89**, 1236 (1985).

28. Y. -M. Tricot, Å. Emeren, and J. H. Fendler, *J. Phys. Chem.*, **89**, 4721 (1985).

29. Y. -M. Tricot and J. H. Fendler, *J. Phys. Chem.*, **90**, 3369 (1986).

30. S. Baral and J. H. Fendler, *J. Am. Chem. Soc.*, **111**, 1604 (1989).

31. X. K. Zhao, S. Baral, and J. H. Fendler, *J. Phys. Chem.*, **94**, 2043 (1990).

32. K. C. Yi and J. H. Fendler, *Langmuir* (submitted for publication as a Letter).

33. S. Xu, X. K. Zhao, and J. H. Fendler, *Advanced Materials*, **2**, 183 (1990).

34. H. Watzke and J. H. Fendler, *J. Phys. Chem.*, **91**, 854 (1987).

35. X. K. Zhao, Y. Yuan, and J. H. Fendler, *Langmuir* (in press).

36. R. Rolandi, R. Paradiso, S. Q. Xu, C. Palmer, and J. H. Fendler, *J. Am. Chem. Soc.*, **111**, 5233 (1989).

37. L. Spankel, M. Haase, H. Weller, and A. Henglein, *J. Am. Chem. Soc.*, **109**, 5649 (1987).

38. A. Zunger and A. J. Freeman, *Phys. Rev.*, **B17**, 4850 (1978).

39. Y. Wang, A. Suna, W. Mahler, and R. Kasowski, *J. Chem. Phys.*, **87**, 7315 (1987).

40. J. -M. Zen, F. -R. F. Fan, G. Chen, and A. J. Bard, *Langmuir*, **5**, 1355 (1989).

THE MACROMOLECULAR DESIGN OF SPIDERS' SILKS

John Gosline, Cynthis Nichols, Paul Guerette, Audrey Cheng, and Steve Katz

Department of Zoology, University of British Columbia
Vancouver, B. C., V6T 1Z4, CANADA

Spiders produce a variety of high performance, tensile fibers (silks) by spinning from concentrated, aqueous protein solutions. Recent developments in genetic engineering suggest that the synthesis of structural proteins in microbial culture to produce polymers for fibers may become commercially viable in the future. We believe that information on the structure-property relationships of spider silks, as well as an understanding of the methods employed by spiders to process protein solutions into fibers, may provide important guidance for these developments. We have been studying the structure-property relationships of the dragline and viscid silks from an orb-web spinning spider. The dragline silk represents a high stiffness fiber, with an initial tensile modulus of 10 GPa and a strength of about 1 GPa, whereas viscid silk is elastomeric, with an initial stiffness of about 1 MPa and an extension to rupture of about 200%. Interestingly, both silks have very similar amino acid composition, so these differences likely represent either major differences in monomer sequence or radical differences in fiber formation. Our studies have revealed that the potential for crystallization in these two silks is very similar, but that the viscid silk contains a non-volatile, water soluble component that inhibits crystallization during fiber formation. This component(s) may provide a useful tool for controlling protein crystallization during spinning of synthetic proteins.

1.0 INTRODUCTION

Silks, which can best be defined as externally spun fibrous protein secretions, are produced by an enormous range of insects and related animal groups,[1] but only two species, the moths, *Bombyx mori* and *Antheraea pernyi*, are grown in mass culture for the production of textile fibers from their cocoon silks. Thus, a vast diversity of protein fibers remains unutilized at this time because it has not been economically feasible to raise other species in mass culture. Current developments in genetic engineering, however, make it likely that silk-like proteins can be produced in microbial culture and spun into textile fibers and other structures, but before this vision becomes reality it will be necessary to develop a much more complete understanding of the relationship between gene sequence, protein structure and mechanical properties, as well as an understanding of the mechanisms needed to spin proteins into textile fibers. Spiders provide an ideal biological system for learning about structure-property relationships in protein-based fibrous biopolymers.

All spiders produce silks, but the orb-web spinning Araneid and Uloborid spiders are particularly versatile in that they spin as many as seven different types of silk. Each type of silk is produced by a different gland-spinneret complex and has a very different biological function. Presumably, each of the seven silks has a different chemical structure that is responsible for producing the mechanical properties that are characteristic of the silk. We have been studying several silks from the spider, *Araneus diadematus*, concentrating on the major ampullate gland silk that forms the spider's dragline and the frame of its orb web, as well as the viscid silk that forms the sticky spiral of the web. The viscid silk has two components; the fiber itself is produced by the flagelliform glands, and the glue coating is produced by the aggregate glands. The dragline and viscid silks are of particular interest because they likely represent two extremes of the kinds of properties that can be achieved by silks.

The studies of Denny[2,3] showed that dragline silks have a set of properties that are very similar to those of drawn Nylons, with high stiffness ($E_{init} \approx 10$ GPa) and high strength ($\sigma_{max} \approx 1$ GPa). More important, however, is their extreme toughness, with the energy to break reaching 150 MJm^{-3}. Indeed, it is likely that the ability of these silks to sustain extensions of the order of 35% prior to rupture and to absorb the kinetic energy from the impact of a flying insects hitting the orb-web is one of the most important functional attributes of this silk.[2,4] Viscid silks, on the other hand, are elastomeric, with initial stiffness of about 1 MPa and extensions to rupture ranging from 230%[3] up to 1300%.[5] Thus, these two silks span the range of properties provided by Nylon, Lycra and beyond. To date, no one has found a silk with the extreme stiffness and strength characteristic of super fibers like Kevlar, but then very few of the vast number of different silks have been studied in any detail.

The mechanical properties of silk obviously arise from their molecular structures, and silks have long been known to contain extensive regions of crystalline organization, with the dominant crystal form being that of ß-pleated sheet crystals arranged parallel to the long axis of the silk fiber.[1] Although the crystalline organization has been well characterized by X-ray diffraction,[6] little is known about the non-crystalline regions that exist between crystal domains. X-ray diffraction indicates that 40 to 50% of the protein in *Bombyx mori* cocoon silk is crystalline;[7] thus, the majority of the structure is 'amorphous', by the criterion of X-ray diffraction. Because some silks (e.g. spider's viscid silk) can exhibit long-range elasticity and because even the 'rigid' dragline silks exhibit large extensions to rupture, the role of these amorphous regions may be crucial to structure-property relationships. Unfortunately, X-ray diffraction provides no information on the structure of the amorphous regions.

Amino acid sequence information may provide insights into the molecular structure, but we know very little about silk sequences. The cocoon silk from *Bombyx mori* is known to contain extensive repeats of the crystal-forming hexapeptide (gly-ala-gly-ala-gly-ser) alternating with "amorphous" domains.[1,8] Recently, sequences for two distinct protein components produced by the major ampullate gland of the orb-web spinning spider, *Nephila clavipes*, have been elucidated.[9-11] Spidroin 1 contains multiple copies of a 34 amino acid long repeat sequence motif (Table-I) with an alanine-rich block about 13 residues long and the remainder containing a tripeptide repeat, gly-gly-X. The overall repeat sequence is represented as $(\text{gly-gly-X})_n$-$(\text{ala})_n$ or $(\text{gly-X-gly})_n$-$(\text{A})_n$, X being glutamine, tyrosine or leucine. The second component, Spidroin 2, contains a similar repeating sequence motif (Table-I) with a 10 residue long poly-alanine block and the

Table-I. Consensus repeat sequences for proteins synthesized by the major ampullate gland of *Nephila clavipes*, which are thought to be the major components of the dragline silk (Ref. 11). The underlined regions likely form the ß-sheet crystals. The letters in the sequences below are based on the one letter codes for amino acids, in which A is alanine, G is glycine, L is leucine, N is asparagine, Q is glutamine, R is arginine, S is serine, and Y is tyrosine.

SPIDROIN 1:	AGRGGLGGQGAGAAAAAAAGGAGQGGYGGLGNQG
SPIDROIN 2:	GPGQQGPGGYGPGQQGPSGPGSAAAAAAAAAAGPGGYGPGQQGPGGY

remainder containing glycine and proline-rich pentapeptide repeats. Spidroin 1 and Spidroin 2 are believed to be the only two major protein subunits comprising *N. clavipes* dragline silk,[11] but their relative proportions are not at present known. Proline, however, is found only in Spidroin 2, making up about 16% of its composition, and the proline content of *N. clavipes* dragline is 3.5%. This suggests that Spidroin 1 makes up roughly 80% of the protein in the fiber. Interestingly, the dragline silk of *Araneus diadematus*, which we have studied, contains about 16% proline (Table-II), and this suggests that *Araneus* silk is very rich in Spidroin 2, or contains other similar proline-rich sequences.

It is possible to infer protein secondary structure from primary sequence, and preliminary studies assign the poly-alanine blocks to the ß-sheet forming portion of the protein. This is based on an observed inter-sheet spacing of 5.3 Å from X-ray diffraction studies of *Nephila* dragline silk, a spacing identical to that of synthetic poly-L-alanine peptides in their ß-sheet conformation.[11,13] It has been suggested[11] that the polyalanine regions are present in an α-helical conformation when the silk is in its aqueous form within the gland. When the silk secretion is subjected to shear forces and mechanical extension during processing, the helical poly-alanine segments likely undergo a helix to ß-sheet transition, forming the ß-crystals that stabilize and reinforce the silk network. The gly-

Table-II. Summary of the amino acid composition of the dragline and viscid silks of the spider *Araneus diadematus*, based on the data of Andersen (Ref.12). The composition values are ex-pressed as residues per 1000, but the values for each silk should not add up to 1000 because individual amino acids may be listed in more than one category. For example, alanine and proline are both listed individually and are also included in the non-polar amino acids.

AMINO ACID	DRAGLINE SILK		VISCID SILK	
Glycine	372	622	442	556
Alanine	176		83	
Serine	74		31	
Proline	158		205	
Acidic	125		55	
Basic	11		31	
Non-Polar	370		390	
Aromatic	41		37	

cine rich regions are similar to sequences found in rubber-like proteins such as elastin,[14] and are designated as portions forming the amorphous poly-peptide chains which are interspersed among the crystalline regions. Interestingly, X-ray diffraction studies of *Araneus diadematus* dragline indicates a larger inter-sheet spacing of about 7.5 Å,[6] which suggests that the crystal regions in this dragline silk likely contain amino acids with side chains that are considerably larger than the single methyl group of alanine. Clearly, much remains to be learned about the details of sequence design in silk proteins from even closely related animals.

While the elucidation of primary structure for other silks will provide insights into possible secondary structural conformation, it can not provide answers to all questions about the organization of the crystalline and amorphous regions. Mechanical and structural characterization of silks provides a different approach to understanding the design of these materials. Previous work[15] demonstrated that dragline silks undergo a dramatic contraction (called a super-contraction) when they are immersed in water, and thermoelastic studies carried out in our laboratory[16] indicate that this hydration-induced contraction produces an elastomeric material in which elastic recoil is driven by polymer conformational entropy changes. This means that the hydrated, contracted silk can be described as a crosslinked network of kinetically-free, random-coil molecules and that its structure can be evaluated with the Kinetic Theory of Rubber Elasticity.[17] For the past several years we have been studying the force-elongation behavior and the strain-birefringence of spider silks in their elastomeric states, and we have used rubber network theory to develop a molecular model for these materials. This approach has allowed us to make estimates of the dimensions of the crystalline and amorphous components in the network, and has provided insights into the process of fiber formation.

2.0 EXPERIMENTAL APPROACH

When dragline silk is immersed in water it contracts to about 60% of its initial length, absorbs water, swells to about twice its dry volume and shows a dramatic change in mechanical properties. The change in mechanical properties is perhaps best indicated by the change in initial tensile modulus, which falls by a factor of about 1000. Because spider silk fibers are usually only microns in diameter, the forces required to test silk samples in the hydrated, elastomeric state are extremely small, and we have developed

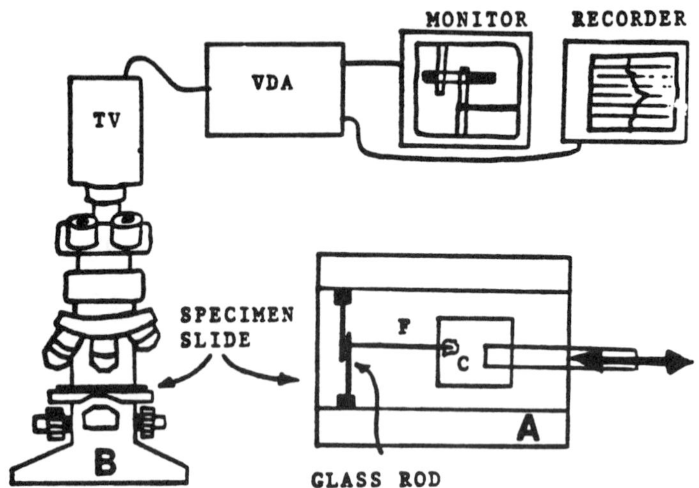

Figure 1. Diagram of the apparatus used to determine the mechanical and optical properties of silk fibers. A. Detail of the specimen slide; the test fiber (F) is attached at one end to a glass rod and at the other end to a movable cover slide (C). B. Microscope system: the specimen slide is mounted on the microscope, and a TV camera (TV) and monitor are used to follow the deflection of the glass rod. A video dimension analyzer (VDA) generates an electronic window that can follow the movement of the contrast boundary created by the edge of the glass rod. It provides a voltage output that is proportional to the movement of the rod.

a microscope-based test system that employs thin glass rods (typically 100 μm in diameter and 1-2 cm long) as force sensing elements. These glass rods are mounted on microscope slides, and force is determined by observing the deflection of the glass rods with a video measurement device attached to the microscope Figure 1. We have used this system to measure micro-Newton forces, and it allows us to determine accurately the shape of the force-elongation curves of dragline and viscid silks in their elastomeric states. In addition, the test system is built on a polarizing microscope, so it is possible to determine the birefringence both in the resting state and at various levels of extension. Because the fibers are extremely thin, we measure the intensity changes created by the measuring compensator with a photo-multiplier that samples the light intensity in the image plane of the microscope's objective lens through a 50 μm diameter light guide. With a 75X objective lens the system samples a 0.67 μm diameter circle that is placed at the center of the silk fiber.

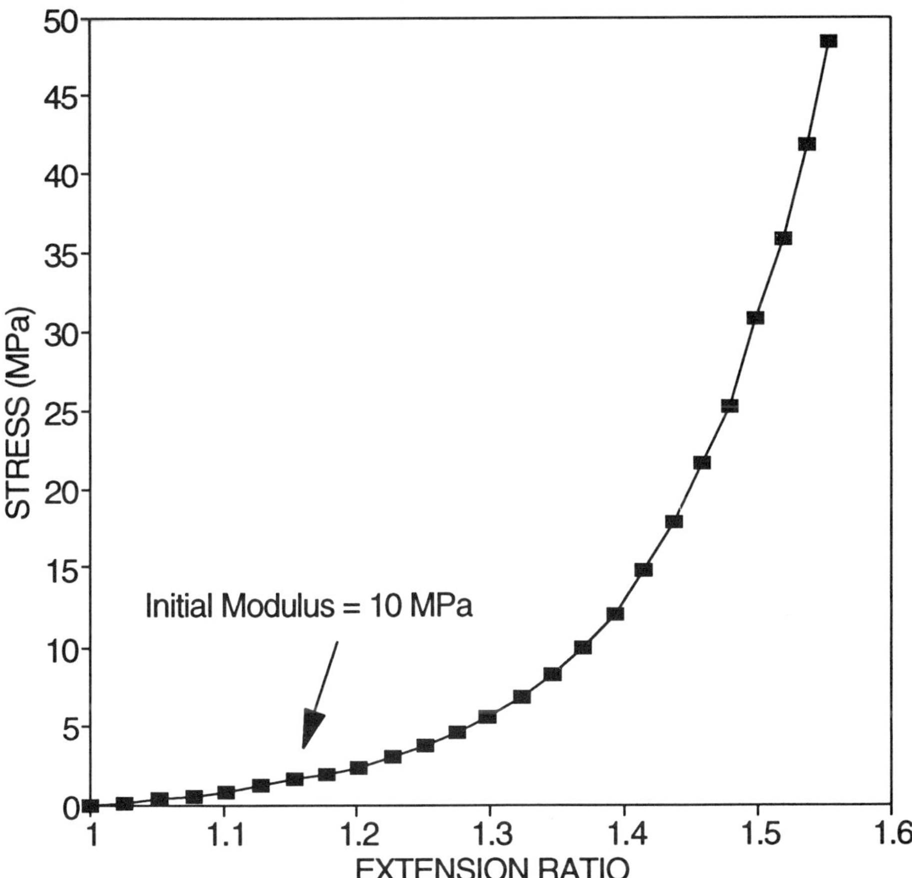

Figure 2. Typical stress-elongation curve for hydrated, contracted dragline silk. Extension ratio is the stretched length divided by the unstretched length; thus, extension ratio 1.5 is a 50% extension.

Figure 2 shows a typical stress-elongation curve for hydrated dragline silk. The most striking feature of this curve is its extreme non-linearity. The stiffness (i.e. slope of the curve) rises roughly 100 fold over the first 75% of extension. In contrast, most lightly crosslinked rubbers exhibit roughly constant stiffness over this range of extensions. Such non-linear stress-elongation behavior is observed in polymer networks only when stretching causes the separation between network junction points to approach the ex-

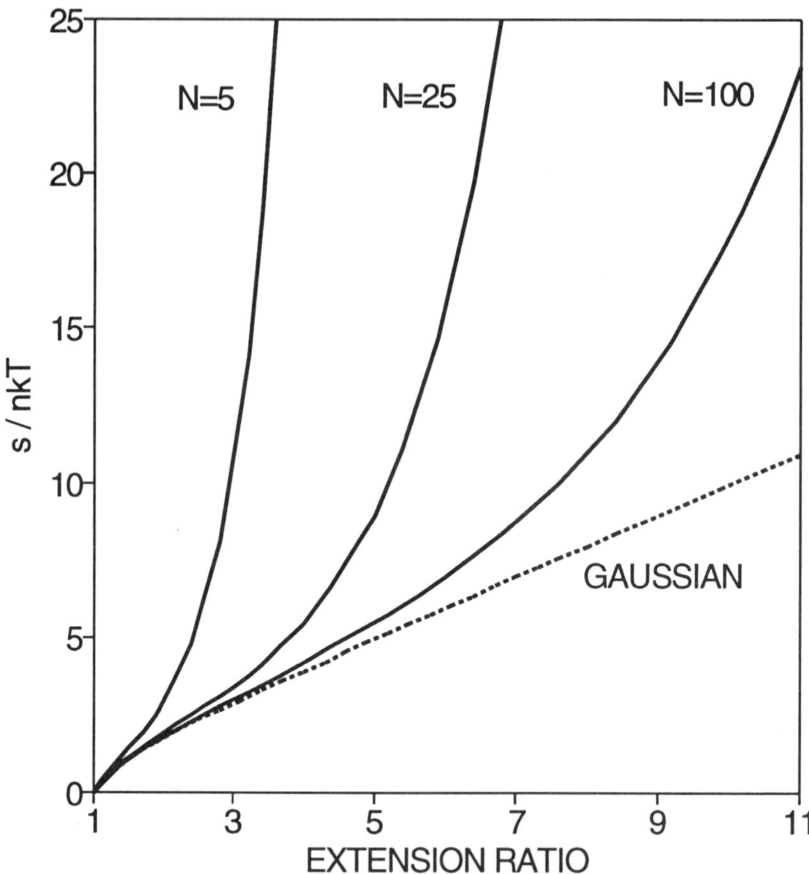

Figure 3. Stress-elongation behavior expected for random polymer networks with different chain lengths, where chain length is given by the number of equivalent random links, N. The stress, s, expressed as s/nkT, is a unitless measure of the force per chain in the network.

tended length of the polymer chains, and this behavior can be analyzed with non-Gaussian network models.[18]

The polymer chains in a random network are modeled as random-walks of N steps (or as random-coil molecules of N equivalent random links), and Figure 3 shows the stress-elongation behavior expected for polymer networks with different chain lengths, N. Note that the stress-elongation behavior departs from the Gaussian curve (broken line) only at large extensions for long-chain networks where N is large, but at small extensions for short-chain networks in which N is small. Also note that stress in this figure is expressed as a unitless ratio, s/nkT, where s is the force divided by initial cross-sectional area, n is the number of network chains per unit volume, k is the Boltzmann constant and T is absolute temperature. According to the kinetic theory, the stiffness, or specifically the shear modulus, G = nkT. Further, if we know the density of random chains in the network (ρ, expressed as kg polymer per m^3 of hydrated network) it is possible to estimate the average molecular weight of the polymer chains between junction points in the network, M_c, as

$$M_c = \rho RT / G \tag{1}$$

where R is the gas constant. Thus, a knowledge of the shear modulus, G, and the shape of the stress-elongation curve will provide a measure of the size of the polymer chains in the network and of the number of equivalent random links that make up each of these polymer chains.

3.0 RESULTS

3.1 HYDRATED, CONTRACTED DRAGLINE SILK

The comparison between Figures 2 and 3 suggests that dragline silk contains a short chain network, in which N, the number of random links per random chain, is small. The initial tensile modulus (E_{init}) is about 10 MPa, and assuming that the shear modulus $G = E_{init}/3$, we can use Eq. 1 to estimate that M_c, the molecular weight of network chains, is approximately 0.45 kD. This calculation is based on a density, ρ = 625 kg m^{-3}, which reflects the fact that hydrated silk is approximately 50% water by volume. On the basis of its amino acid composition,[12] we calculate the average amino acid weight to be 85 g $mole^{-1}$. Thus, from Eq. 1, each network chain should be about 5 amino acids long.

This analysis is, however, suspect for two reasons. First, the silk is clearly a short-chain network and is likely to be well into its non-Gaussian behavior, even for the smallest extensions. Thus, we likely overestimated the magnitude of G. Second, the theory of rubber elasticity is based on networks of random-coil molecules linked together at crosslink sites that do not occupy volume. We know, however, that silks contain extensive regions of crystalline sub-structure that may occupy up to 50% of the total volume. X-ray diffraction analysis of hydrated-contracted dragline silk[19] indicates that the crystals found in the dry, native fiber remain intact when the silk fiber is hydrated. Thus, this elastomeric fiber is in reality a filled-rubber, and it will be necessary to account for the effect of the rigid filler (i.e., crystal) particles on the properties of the system. Filler particles provide a dramatic stiffening effect to rubber networks, and the method used to account for their reinforcing effect is to assume that fillers cause a strain amplification of the network chains that run between the particles.[20] That is, the microscopic strain on a network chain is magnified relative to the macroscopic strain on the whole specimen by the interaction of the network with the rigid filler particles. The magnitude of the strain amplification depends on both the volume fraction and the shape of the filler particles in the system. Round particles give little reinforcement and hence create little strain amplification. Elongated particle shapes provide greater reinforcement and greater strain amplification. In fact, as particles become more elongated, filled rubbers become fiber-reinforced composites. Unfortunately, we have no direct information on the shape or volume fraction of crystal-line particles in this silk, and we must, therefore, resort to an elaborate train of logic to make inferences about the structure of this filled network.

First, application of non-Gaussian network theory allows us to estimate the value of N, the number of equivalent random links in the network chains, and G, the shear modulus of the network, from the shape of the stress-extension curve. This is done by converting the original extension values into a non-Gaussian extension term, using Treloar's series expansion approximation to the inverse Langevan function for the non-Gaussian behavior of a tetrahedrally coordinated random-network.[18] The value of N that best describes the silk network is obtained by testing a range of values and selecting that value which gives the best, linear relationship with the measured force. In fact, linear regression analysis is employed to determine the value of N that gives the largest value of R^2, the coefficient of determination of the linear regression. Figure 4 shows a typical regression plot giving a best value for N. It can be seen that the fit of the stress-extension curve to the non-Gaussian model is excellent, as indicated by the extremely high value of R^2. The slope of this regression gives a "true" value for G, which can be used with Equation 1 to calculate a value of M_c. The results of these calculations are shown in Table-III. Note, however, that results are presented for several values of

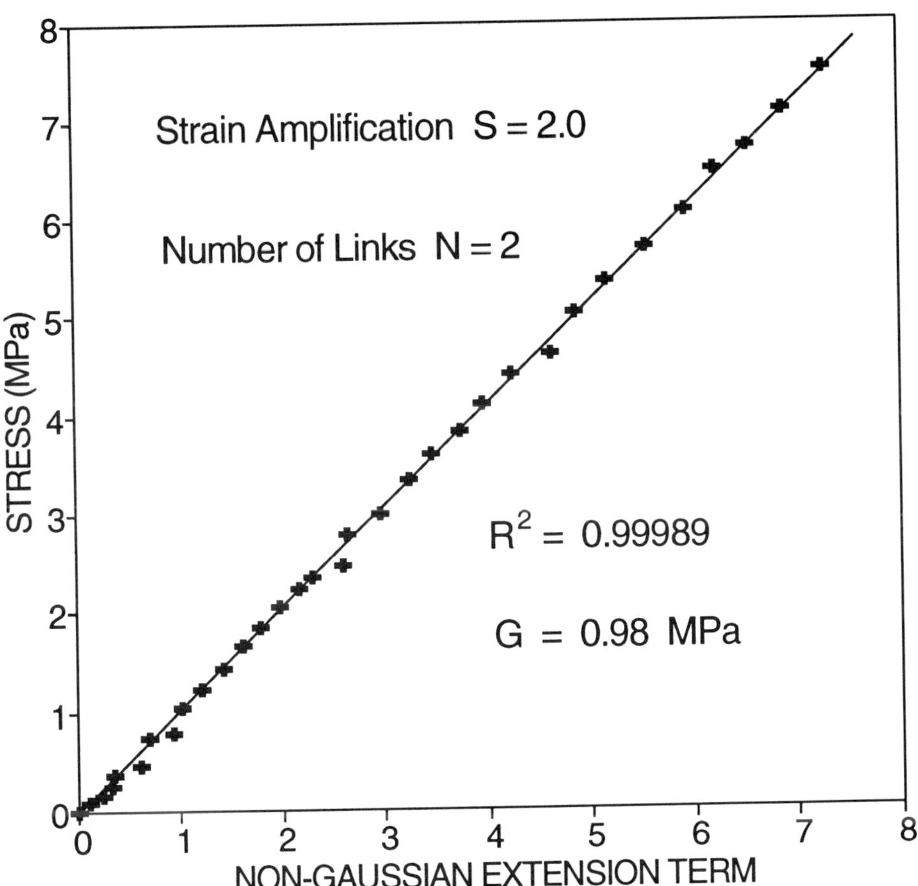

Figure 4. The non-linear stress-elongation curves of hydrated, contracted dragline silk can be fitted to non-Gaussian network curves, as calculated from the series expansion of Treloar (Ref. 18). The slope of the regression line with the greatest R^2 gives the elastic modulus, G.

Table-III. Summary of calculated values from the non-Gaussian analysis of hydrated-contracted dragline silk. For each value of the strain amplification factor, S, regression analysis was used to determine the best estimate of the number of equivalent random links, N, the apparent elastic modulus, G, the average molecular weight between crosslinks, M_c, and the number of amino acid residues per random link, A , as described in the text.

S	N	G (MPa)	M_c (kD)	A
1.0	0.3	0.006	200	7,800
1.5	1.3	0.48	2.4	22
2.0	2.0	0.98	1.36	8
2.5	2.7	1.06	1.15	5
3.0	3.5	1.10	1.09	3.7

strain amplification, S, because we have no *a priori* method of predicting what S should be. For S values greater than one, the actual extension values are multiplied by the strain amplification prior to the analysis described above.

The question now is which value of S is the appropriate one. The criterion we have chosen is the number of amino acid residues, A, that together provide sufficient flexibility to the protein backbone to allow it to act as a random link in the network of random-coil molecules. On the basis of mechanical and optical studies with two crosslinked protein networks, resilin[21] and elastin,[22] and intrinsic viscosity studies on a range of denatured globular proteins[23] we know that A is essentially identical for all proteins, having a value of around 8 to 10 amino acid residues per equivalent random link.

Now, referring to Table-III, at S = 1 (i.e., no amplification or a network with no filler particles) the estimated M_c value is 200 kD, a value that is much larger than our initial estimate of M_c = 0.45 kD. The value of A, however, is about 1000 times too large, so we must look to higher S values to find an appropriate combination of parameters. At S = 2.0 we see that A = 8, and we now have a set of parameters that are at least consistent with what we know about the intrinsic flexibility of polypeptide chains. The results suggest that the protein chains in this filled rubber network behave like random-coil molecules containing two random links. Further, the molecular weight between crosslinks, M_c = 1.35 kD, translates into about 16 amino acids per random chain between crosslinking crystals in the network. This is a very interesting observation, in light of the amino acid sequence data discussed previously. The amino acid repeat sequences of Spidroin 1 and 2 (Table-I) contain roughly 20-30 residue long sequence blocks that should not fit into ß-sheet crystals and likely form the amorphous regions of the network in *Nephila* dragline, and this length is not dramatically different from the length of the "effective random chains" inferred in our mechanical analysis *Araneus diadematus* dragline. The lack of a better agreement probably reflects the fact, discussed previously, that *Araneus* and *Nephila* silks likely contain different silk proteins.

Although we have established a firm criterion for selecting an appropriate value for the strain amplification, we are unable to determine a single combination of volume fraction and particle shape that could explain this level of strain amplification. The only data that bear on this question are some electron microscopic observations of the crystal domains in the cocoon silk of the silk moth, *Bombyx mori.*[1] The crystals are estimated to be rod-like structures with thickness ≈ 2.5 nm, width ≈ 6 nm and length ≈ 20 nm. Considering the uncertainties associated with these values and the fact that they describe a silk that is known to have a distinctly different sequence structure, it is likely that a guess at an elongated filler with an axial ratio of about 5:1 is a reasonable start. Filler particles with this axial ratio will create a strain amplification of S = 2 if they occupy approximately 12% of the network volume.[20] Thus, our current model of the molecular network in hydrated, contracted dragline silk contains a random network with chains of approximately 16 amino acids in length connecting rigid crystals with an axial ratio of about 5:1 and occupying about 12% of the volume of the network.

We have attempted to test this model by predicting the pattern of change in birefringence that would occur if this network is extended. Non-Gaussian network theory allows us to predict the strain birefringence, B, associated with the extension of a network of chains, each with two random links, that occupy 88% of the network volume. The predicted network birefringence is shown as the solid line in Figure 5, and the symbols show individual measurements of the birefringence of a single silk sample taken at the extensions indicated. If our model for the hydrated silk is correct, then the difference between the predicted network birefringence and the measured birefringence, ΔB , should reflect the contribution from the crystals, and the change in ΔB should reflect the alignment of these crystals with the fiber axis during extension.

The unstretched fiber has a birefringence, B = 0.004, which indicates that the crystals are not randomly oriented in the unstretched network, but retain a degree of preferred orientation parallel to the fiber axis. Axial extension of this network should cause the crystals to become more highly aligned with the fiber axis, and at some point the crystals should become maximally aligned so that continued extension does not increase their alignment any further. As shown in Figure 6, which is a plot of ΔB against fiber extension, this is exactly what the results seem to show. Thus, the network model derived from the mechanical analysis seems consistent with the independent optical data, and this gives us some confidence in the validity of the overall model.

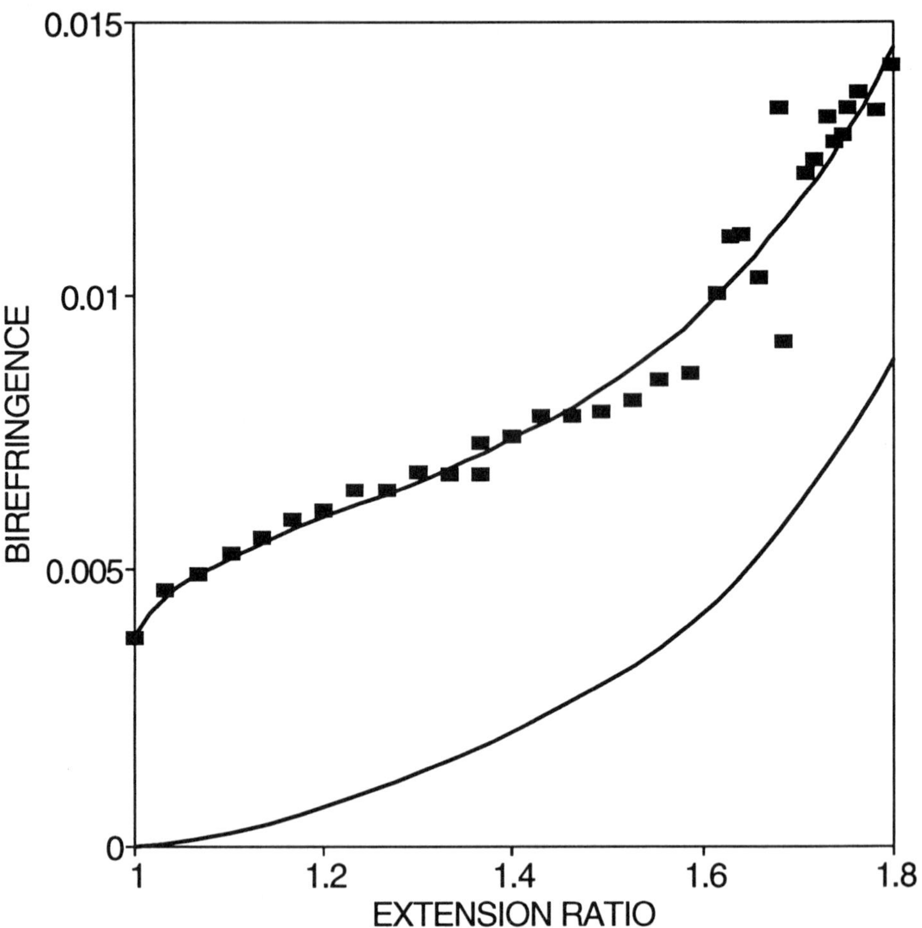

Figure 5. The strain-birefringence of hydrated, contracted dragline silk. The solid line gives the strain-birefringence expected for a random polymer network, in which each chain has two links, that occupy 88% of the network volume. The square symbols are observed birefringence values for a single dragline silk fiber, which are fitted to a third order polynomial.

Finally, we should translate this model for the hydrated, contracted silk back to the extended and dried state that the spider actually uses for its dragline and web frame. To bring the contracted silk back to its initial length, it must be stretched by about 70%. This will have the effect of aligning all of the crystals with the fiber axis, and in addition all the network chains will have been extended far into the non-Gaussian region and will also be virtually parallel with the fiber axis. Now, if we remove the water,

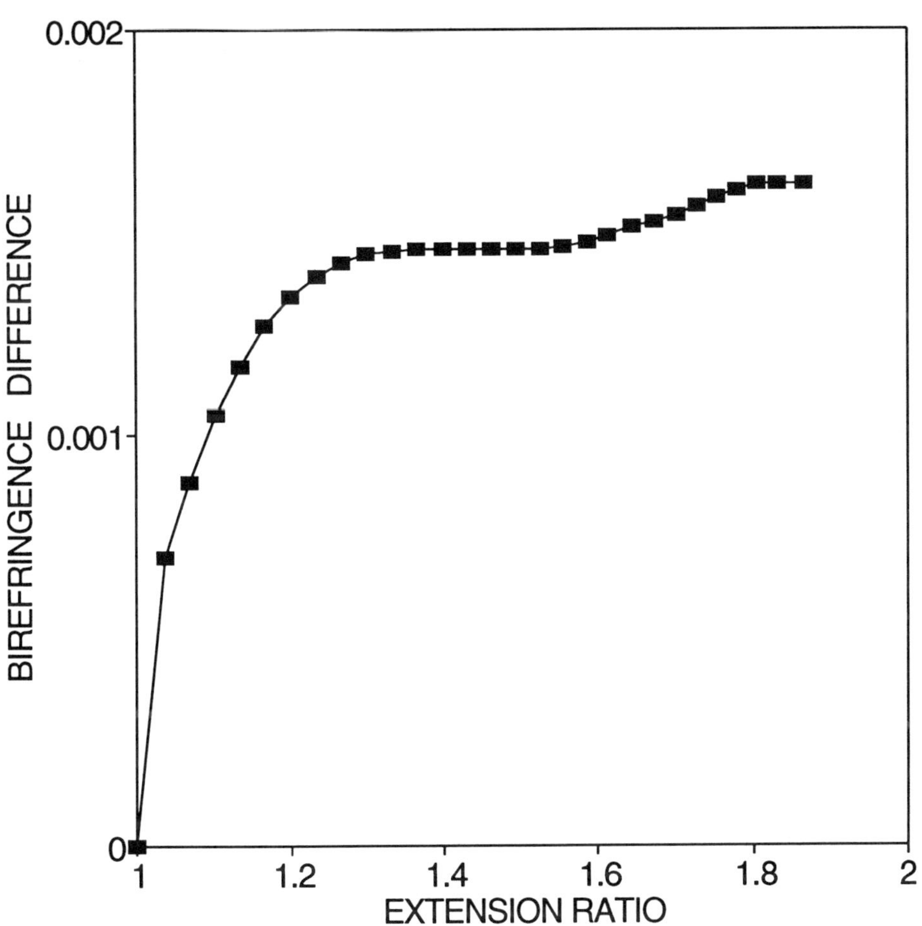

Figure 6. A plot of the difference in the birefringence measured for the silk fiber and the birefringence predicted for the random network component of the fiber. This difference, ΔB, should reflect the alignment of the crystal regions as the silk fiber is extended. The trend suggests that crystals become fully aligned with the fiber axis by an extension of about 40%. The rise in ΔB between 60 and 80% extension may reflect a strain-induced crystallization.

which accounts for half the volume of the hydrated fiber, we will create a dry fiber in which the crystals occupy roughly 25% of the total volume. In addition, the drying will shift the kinetically-free chains in the amorphous network well beyond the glass transition and lock them into this extended and highly oriented structure. Thus, the

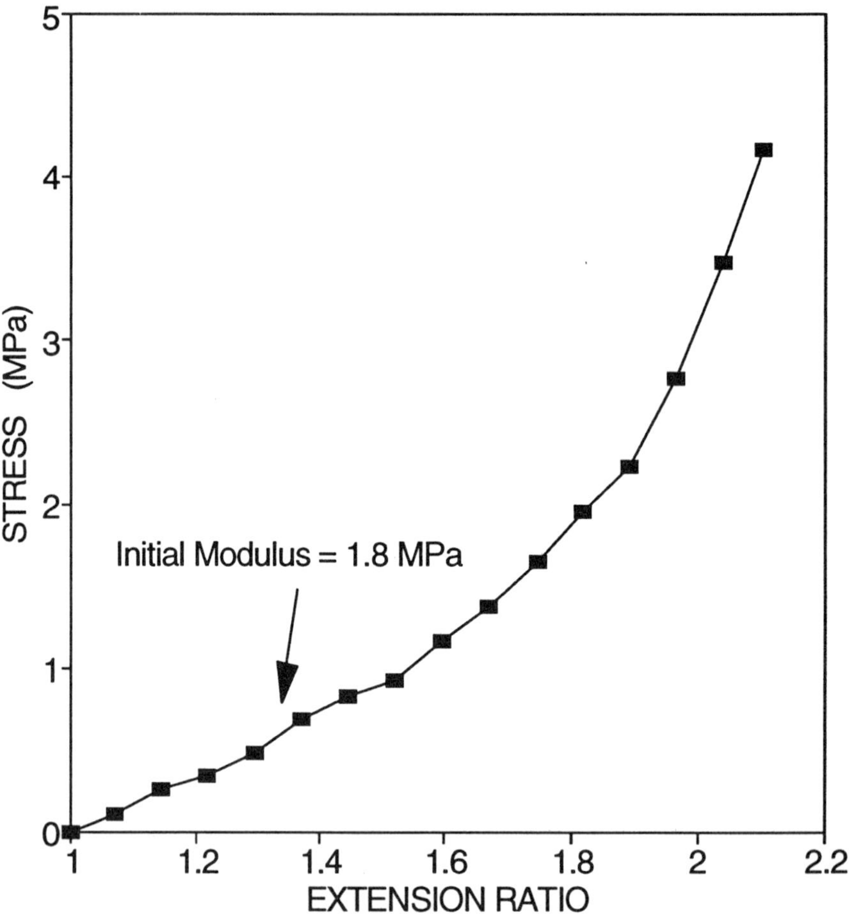

Figure 7. Typical stress-elongation data for native viscid silk taken directly from the web with its glue layer intact.

extension of the silk, which probably occurs in nature as the spider draws its silk out of the spinneret under its own body weight, creates the molecular orientation that is responsible for the high stiffness and strength of the dry silk.

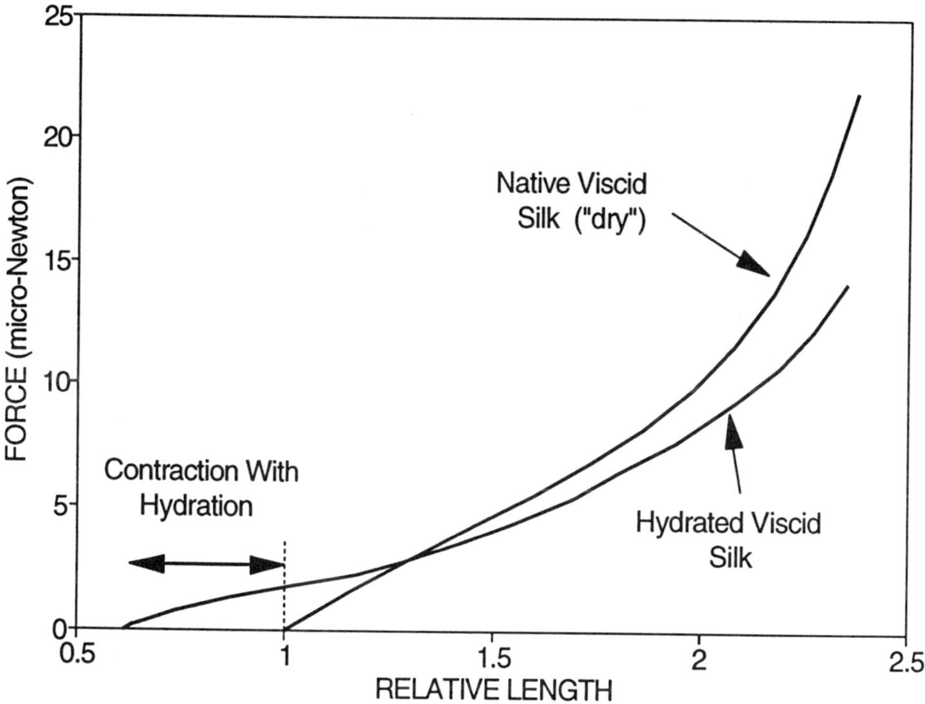

Figure 8. Force-elongation behavior for native viscid silk, and for the same sample when immersed in water.

3.2 VISCID SILK

Figure 7 shows typical stress-elongation behavior for native viscid silk, taken directly from the web, with its glue layer still intact. Note that the mechanical behavior of this "dry", native fiber is very similar to that of dragline silk when it is immersed in water. The initial stiffness of the native viscid silk is about 2 M Pa, and the fiber can be stretched to more than twice its initial length before it breaks. In other words, viscid silk is elastomeric in its native state.

When viscid silk is immersed in water it, like dragline silk, exhibits a dramatic contraction and swelling, as shown in Figure 8, but because the native silk is already elastomeric, the hydration of viscid silk does not cause such a dramatic shift in mechanical properties. In the example shown in Figure 8 hydration caused this sample of viscid silk to contract to about 60% of its native length, but there was only a small decrease in the initial slope of the force-extension curve. However, the cross-sectional area of the contracted fiber increased by almost two fold, causing the initial modulus of the hydrated viscid silk to fall to about one third that of the native fiber. The magnitude of the contraction and the fall in stiffness of viscid silk is quite variable (the contraction ratio ranges from 0.4 to 0.85, and the fall in initial stiffness can vary between 2 and 4 fold), suggesting a large variation in the degree to which hydration affects the crosslinking of the viscid silk network. At this time we do not completely understand how or why water affects the network crosslinks. It appears, however, that viscid silk, like dragline silk, is crosslinked by crystal domains or other non-covalent, physical interactions because treatment of viscid silk with non-hydrolytic, protein denaturing agents, such as 6 M guanidine hydrochloride, causes the silk to dissolve. Thus, the mechanical behavior suggests that viscid silk can most likely be modeled as a lightly crosslinked rubber network, in which the crosslinks are formed at least in part by crystalline domains.

Polarized light studies indicate that unstretched, native viscid silk has a positive birefringence of about 10^{-3}, indicating either that there are relatively few crystals or that they are not highly oriented. Extension, however, causes an increase in birefringence, with a stress-optical coefficient of approximately $2 \times 10^{-9}\ m^2\ N^{-1}$, a value that is roughly twice that seen for the hydrated protein networks found in resilin[21] and elastin.[22] Resilin and elastin, however, are crosslinked by covalent interactions between amino acid side chains and are virtually totally amorphous. The large stress-optical coefficient seen for viscid silk likely reflects the high polarizibility of the crystal domains that form its crosslinks.

It is not clear, however, if the crystal domains in viscid silk are sufficiently large or abundant to act as reinforcing fillers. The shear modulus of this silk, based on the initial slope of the stress-elongation curve ($G = E_{init}/3$), is approximately 0.6 MPa. It is difficult to determine what the density, ρ, of the polymer is in the native silk strand, but we estimate that it is somewhat higher than that of fully hydrated frame silk. Assuming $\rho \approx 850\ kg\ m^{-3}$, M_c is approximately 3.5 kD. The shape of the stress-elongation curve of native viscid silk suggests that N, the number of links per chain, can range from 3 to 10, with an average value of 4.7. For an average residue weight of 85 g mole^{-1} and an equivalent random link containing 8 - 10 amino acids, M_c should be between 2 and 8.5 kD, with an average value of around 3.6 kD. This close match between the two estimates of M_c suggests that there is no need to invoke the filler particle-strain amplification argument used for the much more crystalline dragline silk.

3.3 CRYSTALLINE VERSUS NON-CRYSTALLINE SILKS

This comparison of dragline and viscid silks clearly indicates that spiders are capable of producing a broad range of mechanical properties in the fibers that they spin, and this leads us to the question of how they modulate these properties. One possibility is that the differences in crystallinity between viscid and dragline silk lie primarily in the chemical sequence of the proteins, but we do not have enough information to make even tentative conclusions because there is no information on the sequence of viscid silk. We must, therefore, rely on amino acid composition data (Table-II) as an indication of chemical structure. One might expect the viscid silk composition to be markedly different from that of the dragline silk, but there is a strong similarity between the two compositions. Both silks contain a large fraction of glycine, alanine, and serine, which are the key amino acids in the crystals of *Bombyx mori* silk (Note: these three amino acids make up more than 85% of the silkworm's cocoon silk fibroin.) There is, however, less alanine in viscid silk than in dragline silk, but the large inter-sheet spacing for *Araneus* dragline crystals seems to suggest that alanine is not such a key parameter in crystal formation as in the case of *Nephila* dragline. Thus, there is no compelling trend to suggest that viscid silk has a dramatically reduced potential for forming crystals. Viscid silk has a very high proline content, which may serve to inhibit crystal formation in regions of the sequence. But dragline silk also has quite a high proline content. What is clear, therefore, is that it is not possible to make any firm conclusions about the capacity of these two proteins to crystallize from a knowledge of their amino acid composition, and we must wait for definitive sequence data for the two silks. One might suspect, however, considering the similarity in composition and the extreme difference in physical properties, that there are some very interesting differences in sequence structure.

A recent article[24] suggests that the mechanical properties of the viscid silk are modulated by the presence of the glue rather than by differences in protein sequence, and in particular it is suggested that the water in the glue maintains the viscid silk in its elastomeric form. We have observed similar effects, but we suspect that the controlling

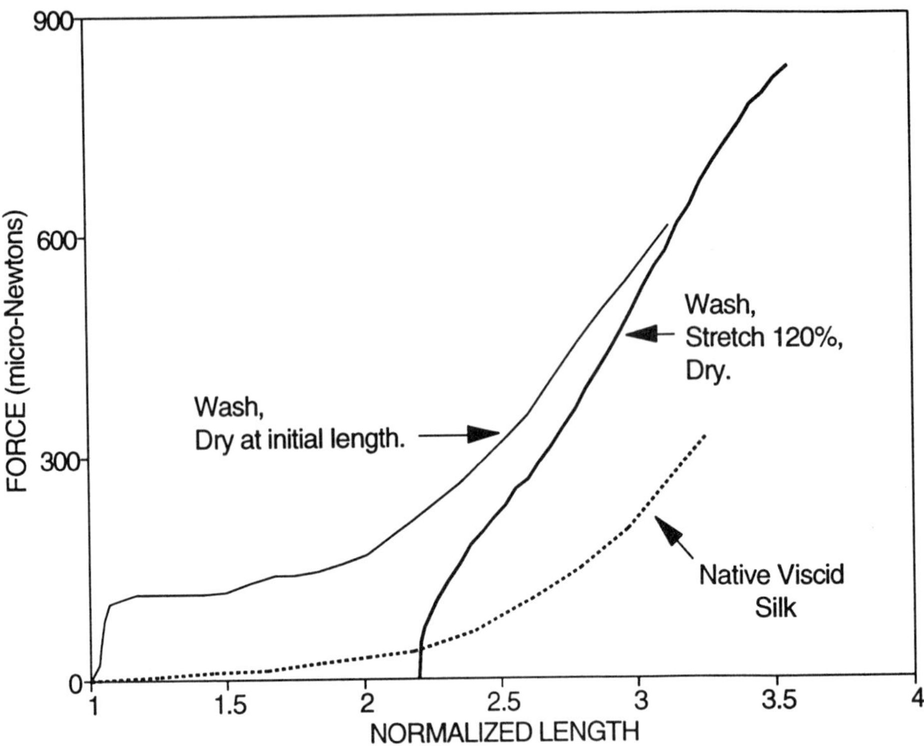

Figure 9. Force-elongation behavior for native viscid silk and for samples that have been "washed" in distilled water to remove water soluble crystal inhibitors from the glue. One washed sample was allowed to dry and crystallize at its initial length, and the other was allowed to dry and crystallize after stretching in water by 120%. Each curve represents a test to failure for a different sample taken from the same web.

factor is not water, but rather some water-soluble, non-volatile compound that is present in the glue (i.e., synthesized by the aggregate gland) or incorporated into the core silk fiber within the flagelliform gland.

Our logic is as follows. Figure 9 shows the force-elongation behavior of viscid silk samples that have been "washed" in distilled water and then dried at various lengths. The initial, reference length for these fibers is the length at which the native fiber becomes slack when it is allowed to recoil quickly. It is important that the recoil be rapid (several seconds) because the surface tension of the glue on native fibers will gradually

Table-IV. Birefringence and fiber diameters measured for viscid silk fibers from the web of *Araneus diadematus*. Native silk in the dry condition means silk strands directly from the web. These native fibers in both the wet and dry conditions exhibit rubber-like elasticity and low crystallinity. "Washed" silks are silks that have been allowed to dry in air (R.H. about 50%) at least once after having been immersed in an excess of distilled water. This washing apparently removes water-soluble, non-volatile compounds from the silk fiber.

Test Condition	Birefringence	Diameter (mm)
I. Native Silk (dry)	1.4×10^{-3}	2.3
II. Native Silk (in water)	0.9×10^{-3}	3.4
III.		
A. "Washed" Silk (dry, $L = L_0$)	1.6×10^{-2}	1.1
B. "Washed" Silk (re-wet, $L = L_0$)	5.0×10^{-3}	1.5
IV.		
A. "Washed" Silk (dry, $L = 1.4\ L_0$)	5.3×10^{-2}	0.8
B. "Washed" Silk (re-wet, $L = 1.4\ L_0$)	1.1×10^{-2}	1.5

take up any slack, and thus, the core fibers can become coiled inside the glue drops. Thus, the initial, reference length represents the unstretched state of the native core fiber. When "washed" fibers are dried at this reference length the fibers have dramatically different properties. Their initial modulus increases by about four orders of magnitude to approximately 10 GPa. Polarized light studies (Table-IV) indicate that the "washed" fibers have become highly birefringent, and this indicates that the viscid silk proteins, like the dragline silk proteins, have a strong tendency for crystal formation, but somehow this crystallization has been inhibited during fiber formation.

Contrary to the previous interpretation,[24] we believe that the factor which prevents crystallization in the native fiber is not the water bound by a hygroscopic glue, but rather some specific compound that inhibits crystallization. The water in the native fiber sitting in air at about 50% relative humidity will experience a water activity determined by that relative humidity. It does not matter if the glue is hygroscopic and can draw water from the air; that water is not available to hydrate the core silk fiber because the hygroscopic glue will draw water from the silk as well. Thus, the core fiber in the native condition remains in its elastomeric, low-crystalline state because there is a crystallization inhibitor present, and this inhibitor must be relatively small in order for it to penetrate the rather dense protein network present in the fiber. Andersen[12] and Vollrath et al.[25] report that the glue of Araneid, viscid silk contains large glyco-proteins as well as several small organic molecules that are present in very high concentrations. The most abundant small compounds include γ-amino-butyramide, choline, isethionic acid, N-acetyl taurine, and betaine. For example, the glue of *Araneus diadematus* contains 1.9 M γ-amino-butyramide and 1.5 M choline, as well as smaller concentrations of the other compounds.[25] We do not know which, if any, or what combination of these compounds functions as the crystal inhibitor, but all are highly polar and are likely able to interact with the peptide groups of the silk polypeptide backbone.

If small molecules, like these, do function to control crystal formation in spider's viscid silk, then perhaps compounds of this nature will prove useful in modulating fiber formation in synthetic processes. Examination of Figure 9 and Table-IV shows some of the possibilities. If fibers can be spun to form lightly crosslinked polymer networks, like native viscid silk, and then drawn to varying degrees in the solvent-swollen state and allowed to crystallize, it will be possible to modulate the degree of molecular orientation and produce a wide range of mechanical properties. In our very preliminary efforts we have produced highly ductile fibers when crystallization occurs at low extensions, and stiff, strong and tough fibers when crystallization occurs at higher extensions. Perhaps this method will allow us to produce Kevlar-like super fibers under appropriate conditions. One of the major problems of applying what we know about spider silks to the design of synthetic systems is the extreme hydration sensitivity of silks. Dragline silk turns to rubber when it gets wet, and the washed, dried viscid silk also contracts and becomes more elastomeric when it is re-hydrated. Note, however, that re-wetting "washed" viscid silk that was crystallized at a 40% extension does not return the fiber to its initial low birefringence condition. Rather, the birefringence remains well above the residual birefringence of hydrated, contracted dragline silk (Figure 5), and this suggests that the viscid silk proteins may form a more stable, oriented network. It is possible that the somewhat higher content of non-polar amino acids in viscid silk explains this difference.

4.0 CONCLUSIONS

Spiders produce a wide range of fibrous biopolymers that provide an interesting set of "model" materials for developing an understanding of the structure-property relationships in protein-based systems. Our investigations of the mechanical and optical properties of these silks are beginning to reveal important aspects of their network structure and may reveal clues to the control of crystallization during fiber formation. Clearly, there is much more to be learned. We need to know more about the sequence design of natural silks before we invest major resources in the construction of artificial genes for the production of synthetic silk-like proteins. In addition, we are largely ignorant about the processes involved in the conversion of the liquid-crystalline primary silk secretion[26] into the finished fiber. Future work with spider silks is likely to make important contributions to these investigations.

5.0 ACKNOWLEDGEMENTS

This project was supported by research grant 586934 from the Canadian Natural Sciences and Engineering Research Council and by a research grant from the E. I DuPont de Nemours & Company.

6.0 REFERENCES

1. F. Lucas and K. M. Rudall, Silks, in *Comprehensive Biochemistry,* M. Florkin and E. Stotz (eds.) **26B** (1968) pp. 475-588.

2. M. W. Denny, "The Physical Properties of Spider's Silk and Their Role in the Design of Orb-Webs," *J. Exp. Biol.,* **65**, 483-506 (1976).

3. M. W. Denny, "Silks, Their Properties and Functions," in *Mech. Properties of Biol. Mat.,* J. F. V. Vincent and J. D. Currey (eds.) *Soc. Exptl. Biol.* (1980) pp. 247-271.

4. J. Gosline, "Efficiency and Other Criteria for Assessing the Quality of Structural Biomaterials," in *Efficiency and Economy in Animal Physiology,* A. Blake (ed.) (Cambridge University Press, 1991).

5. J. DeWilde, "Some Physical Properties of the Spinning Threads of *Araneus diadematus," Arch. Neerl. Sci.*, **27**, 118-132 (1943).

6. F. Lucas and K. M. Rudall, "Extracellular Fibrous Proteins: the Silks," in *Comprehensive Biochemistry*, M. Florkin and E. H. Stotz (eds.) Vol. **26B** (Elsevier, Amsterdam, 1968) pp. 475-558.

7. E. Izuka, "Degree of Crystallinity and Modulus Relationship of Silk Threads from *Bombyx mori* and Other Moths," *Biorheology*, **3**, 1-8 (1965).

8. L. P. Gage and R. F. Manning, "Internal Structure of the Silk Fibroin Gene of *Bombyx mori*. I. The Fibroin Gene Consists of a Homogeneous Alternating Array of Repetitious Crystalline and Amorphous Coding Sequences," *J. Biol. Chem.*, **255**, 9444-9450 (1980).

9. M. Xu and R. V. Lewis, "Structure of a Protein Superfiber: Spider Dragline Silk," *Proc. Natl. Acad. Sci., USA*, **87**, 7120-7124 (1990).

10. M. B. Hinman and R. V. Lewis, "Isolation of a Clone Encoding a Second Dragline Silk Fibroin," *J. Biol. Chem.*, **267**, 19320-19324 (1992).

11. R. V. Lewis, "Spider Silk: the Unravelling of a Mystery," *Acc. Chem. Res.*, **25**, 392-398 (1992).

12. S. O. Andersen, "The Amino Acid Composition of Spider Silks," *Comp. Biochem. Physiol*, **35**, 705-711 (1971).

13. J. M. Gosline, "Structure and Mechanical Properties of Rubber-Like Proteins in Animals," *Rubber Chem. Technol.*, **60**, 417-438 (1987).

14. *Conformation in Fibrous Proteins,* R. D. B. Fraser and T. P MacRae (Academic Press, New York, 1973) pp. 228-232.

15. R. W. Work, "Dimensions, Birefringence and Force-Elongation Behavior of Major and Minor Ampullate Silk Fibers from Orb-Web-Spinning Spiders--the Effect of Wetting on These Properties," *Text. Res. J.*, **47**, 650-652 (1977).

16. J. Gosline, M. Denny, and M. DeMont, "Spider Silk as Rubber," *Nature, Lond.*, **309**, 5512-552 (1984).

17. *The Principles of Polymer Chemistry*, P. Flory (Cornell University Press, Ithaca, New York, 1953).

18. *The Physics of Rubber Elasticity*, L. R. G. Treloar (Clarendon Press, Oxford, 1975).

19. R. W. Work and N. Morosoff, "A Physico-Chemical Study of the Supercontraction of Spider Major Ampullate Silk Fibers," *Text. Res. J.*, **52**, 349-356 (1982).

20. L. Mullins, "Theories of Rubber-like Elasticity and the Behaviour of Filled Rubbers," in *The Mechan. Properties of Biol. Mat.*, J. F. V. Vincent and J. D. Currey (eds.) (Soc. for Exptl. Biology, Cambridge, 1980) pp. 273-288.

21. T. Weis-Fogh, "Molecular Interpretation of the Elasticity of Resilin, a Rubber-like Protein," *J. Mol. Biol.*, **3**, 648-667 (1961).

22. B. B. Aaron and J. M. Gosline, "Elastin as a Random-Network Elastomer: a Mechanical and Optical Analysis of Single Elastin Fibers," *Biopolymers*, **20**, 1247-1260 (1981).

23. S. Lapanje and C. Tanford, "Proteins as Random Coils. IV. Osmotic Pressures, Second Virial Coefficients and Unperturbed Dimensions in 6 M Guanidine Hydrochloride," *J. Am. Chem. Soc.*, **89**, 5030-5034 (1967).

24. F. Vollrath and D. T. Edmonds, "Modulation of the Mechanical Properties of Spider Silk by Coating with Water," *Nature*, **340**, 305-307 (1989).

25. F. Vollrath, W. J. Fairbrother, R. J. P. Williams, E. K. Tillinghast, D. T. Bernstein, K. S. Gallagher, and M. A. Townley, "Compounds in the Droplets of the Orb Spider's Viscid Spiral," *Nature*, **345**, 526-527 (1990).

26. K. Kerkam, C. Viney, D. Kaplan, and S. Lombardi, "Liquid Crystallinity of Natural Silk Secretions," *Nature, Lond.*, **349**, 596-598 (1991).

ROLE OF MOLECULAR GENETICS IN POLYMER MATERIALS SCIENCE

Maurille J. Fournier,[#,*] Thomas L. Mason,[#,*] and David A. Tirrell[§,*]

[#]Department of Biochemistry and Molecular Biology
[§]Department of Polymer Science and Engineering
[*]Program in Molecular and Cellular Biology
Lederle Graduate Research Center, University of Massachusetts
Amherst, Massachusetts 01003, USA

The known and likely contributions of molecular genetics to materials science are discussed. Key benefits include: (i) control of synthesis parameters that is superior to current chemical strategies, and (ii) the potential to genetically engineer proteins that will either facilitate materials synthesis or serve as materials themselves. The facilitating proteins include enzymes that catalyze polymerization or modification of biologically related material, and proteins that mediate organization and crystallization of materials components. Recent progress in genetic engineering of novel protein-based materials is also reviewed.

1.0 INTRODUCTION

As noted throughout this volume, living organisms produce a striking array of structural materials with a wide variety of biological functions. These materials are composed of organic polymers, i.e., protein, carbohydrate, and lipid constituents and inorganic components as well. Some, like elastin, collagen, and the silk fibroins, consist entirely of protein molecules. Others like wood, bone, and shell are of mixed composition, developed, for example, from whole cells or mixes of protein, carbohydrate, and minerals. These structural materials are created by an assortment of different synthesis and assembly schemes and the end products exhibit a remarkable range of chemical and physical properties. While biomaterials have long been of fundamental interest to biologists, it has become clear in recent years that these materials also have important and exciting implications for materials scientists. Many important lessons pertaining to polymer synthesis, structure and function will be derived from studies of natural materials.

Drawing on this knowledge, it will eventually be possible to faithfully generate valuable natural products by new strategies and to create important new classes of materials as well. The new materials may be structural variants of natural materials or they may be designed *de novo* from first principles derived from studies of natural or synthetic materials. Both biological and chemical synthesis can be expected to play imporant roles in the production of these materials. In time, it seems likely that synthesis schemes will evolve that feature combinations of biological and chemical strategies. Biological synthesis will be mediated both by cells engineered for this purpose and biological components functioning in a cell-free environment.

2.0 BENEFITS FROM GENETIC ENGINEERING

Without doubt, genetic engineering provides the single most important vehicle for extending the lessons of biological design and synthesis to materials science. The development of this technology has opened the way for cloning both natural and artificial genes and producing useful recombinant products. Adapting this technology to expand the frontiers of materials science is logical and efforts in this direction are already underway. Advances will predictably come from two areas of molecular genetic investigation, one basic, the other more applied. The basic thrust will involve discovery and characterization of natural processes that can be exploited or mimicked for materials synthesis. The second thrust will involve engineering of organisms and biological components for direct use in preparing materials.

The nature of the first thrust will be to define the mechanisms by which natural materials are synthesized, modified, assembled, and degraded. This acknowledge will be developed through characterizing cloned gene systems both *in vivo* and *in vitro.* Isolated genes will be used to produce enzymes and other proteins for mechanistic studies, studies of gene and enzyme regulation, and for creating new enzyme forms with novel performance properties. Logical aims in enzyme engineering will include improving stability and turnover rates, and modifying specificities so that new substrates can be accommodated.

Genetic engineering not only offers a means for obtaining useful biological materials, it also provides the materials scientist with a means of obtaining polymeric materials of greater purity than have generally been available previously. This capability is a hallmark of the exquisitely controlled machinery used to synthesize natural proteins and nucleic acids. From this perspective alone genetic synthesis has important implications for both fundamental and applied materials research.

3.0 BIOLOGICAL vs CHEMICAL SYNTHESIS

For the foreseeable future, genetic synthesis will be favored over chemical synthesis as the route to producing materials-related proteins and nucleic acids. As summarized in Table-I, genetic synthesis provides much better control over synthesis parameters than is currently possible by existing chemical methods of polymer synthesis. Protein polymers generated *in vivo* from genetic templates will be more pure in size, sequence, and stereochemistry than proteins or other polymeric materials produced chemically. It follows from this that control of hierarchical structure will also be superior.

4.0 PROTEIN ENGINEERING AND MATERIALS SCIENCE

At least three classes of genetically engineered proteins can be expected to benefit materials science. These correspond to (i) enzymes that can be used for *in vivo* or in *vitro* production of materials, (ii) proteins that facilitate organization and crystallization of a materials components, and (iii) proteins that are themselves useful materials.

4.1 GENETICALLY ENGINEERED ENZYMES

A key contribution of recombinant DNA technology will be the production of enzymes for the synthesis of natural and new biomaterials. These enzymes will be utilized both in live, recombinant organisms and in cell-free processes. Applications will include polymerization of natural or unnatural precursors and modification of preformed polymeric materials. The new precursors may be produced by either biological or chemical synthesis, and the same holds for preformed polymers to be modified by enzyme treatment.

Enzymes known to participate in the synthesis of interesting natural materials will understandably receive first attention. Candidates will include enzymes that mediate the polymerization of protein and non-protein biomaterials. An example of an especially

Table-I. Methods of Polymer Synthesis

Method	Features	Comments
Chemical		
Step-growth polymerization	$M_w/M_n = 2$ statistical composition, sequence, stereochemistry	
Chain-growth polymerization (e.g., radical)	$M_w/M_n > 1.5$ statistical composition, sequence, stereochemistry	difficult to prepare blocks, telechelics, etc.
Living-polymerization	$M_w/M_n \geq 1.03$ statistical composition, sequence, stereochemistry	blocks, telechelics, starts, are possible
Ziegler-Natta polymerization	M_w/M_n is large statistical composition, sequence, good stereochemical control	limited set of monomers
Genetic		
In vivo	$M_w/M_n = 1.00$ uniform sequence and composition, stereo-chemically pure	diverse functional groups, e.g., alkyl, aryl, -OH, -CO_2H, -NH_2, - SH, others
Cell-free	in principle, as above	can be combined with chemical synthesis, modification

interesting class of synthetic activities are the enzymes that mediate the synthesis of bacterial polyesters such as the polyhydroxyalkanoates (PHA) made by the *Pseudomonas* bacteria such as *Pseudomonas oleovorens.*[1-3] Also relevant are the enzymes in these same organisms that catalyze the depolymerization of these materials. In principle, recombinant enzymes can be used here and with other synthetic processes to generate both the natural polymers or next-generation variants. For example, in the case of the polyesters it may eventually be feasible to produce novel polymers containing alkane precursors which are utilized poorly or not at all by natural microbes. Similarly, development of mutant variants of the depolymerizing enzymes could provide convenient strategies for specific degradation of biopolymers of this type. Thus, there is the potential to genetically engineer both natural enzymes and new enzyme forms with novel specificities, and to utilize these both *in vivo* or *in vitro* strategies.

In addition to polymerizing and de-polymerizing activities a variety of modifying enzymes can be imagined to prove useful in generating novel materials. These enzymes could be used to modify preformed polymers to achieve new physical or chemical properties. In the case of protein-based polymers, for example, a variety of natural enzymes could be used to effect selective cleavage, methylation, acetylation, phosphorylation, and glycosylation. Enzyme mediated modifications can also be envisioned for other classes of biopolymers.

4.2 ORGANIZING PROTEINS

It seems likely that recombinant proteins can be used to mediate organization and crystallization of materials components. This appears to be the situation for several cases of biomineralization under study and the principles involved here may be adapted for organizing other materials. Although the roles of the matrix proteins implicated in biomineralization are not yet known, several interesting possibilities have been proposed.[4] These include: (i) concentration of ions to achieve supersaturation necessary for crystallization ; (ii) inhibition of mineral deposition; (iii) promoting growth of particular crystal isomorphs; (iv) influencing product shape and strength; and (v) providing a structural framework for crystallization. All of these properties are relevant to materials synthesis. The first advances in this area could reasonably be expected to come from efforts to create recombinant materials with nucleation and template organizing functions.

Proteins have the greatest potential for use as templates at this time, because of the diversity of amino acid sequences and folding domains available and our understanding of protein structure. This potential is enhanced by the capability, in principle, of engineering protein polymers of any chosen sequence and controlling peptide folding with high precision. There is, however, no *a priori* reason that precludes the use of nucleic acid, polysaccharide, and lipid polymers as organizing materials. DNA and RNA are precise genetic templates, but the potential of these and other biopolymers to mediate materials assembly is less apparent at this time. Nucleic acid polymers of controlled length and sequence can now be engineered at will by chemical synthesis and recombinant DNA technology. However, this is not yet the case for polysaccharides and lipid polyesters. There is, of course, less sequence versatility with DNA and RNA than with protein and the potential coding capacity of carbohydrate and lipid polymers is quite unknown. While much is known about the crystalline structures of proteins and nucleic acids, methods for analyzing and controlling the molecular architecture of carbohydrate and lipid polymers is much less advanced.

The template activity of a protein will be encoded in the amino acid sequence, but manifest through the folding domains that arise from the sequence. The relevant secondary structural elements include: (i) α-helices, (ii) β-strands, and (iii) β-turns. Interaction of these elements provides the basis for tertiary and quarternary interactions of natural proteins and the binding of natural proteins with other molecules. Template organizing function will be determined by the same chemical and structural properties. It follows then that precision in template function will require detailed understanding and control of peptide chain architecture, to achieve appropriate geometry and interacting forces. As knowledge about three dimensional protein structure and structural dynamics advances it should become feasible to design protein polymers with a wide range of exquisite template functions.

4.3 PROTEIN-BASED MATERIALS

Several proteins are known to have outstanding materials properties. Particularly striking examples are the silks, elastin, collagen, and the marine bioadhesives produced by mussels and barnacles. These particular materials are extracellular matrix proteins with repetitive amino acid sequence elements that are key to their unique properties. Studies with isolated natural proteins are now being augmented with efforts to produce these materials by both chemical and molecular genetic strategies (reviewed in Reference 5 and see below).

In addition to natural proteins it is clear that unnatural protein materials will also play an important role in materials development. Artificial derivatives of natural proteins are currently under study and this area will continue to be a major focus of attention in the future. In addition, completely new materials are being designed from first principles, with natural protein structures serving as paradigms. Genetic synthesis of new protein materials is the main focus of our own program and we return to this concept below.

5.0 ADVANCES WITH PROTEIN-BASED MATERIALS

5.1 VARIANTS OF NATURAL PROTEINS

A variety of natural silk proteins are being characterized and genetic engineering of silk fibroin variants is in development.[6-12] Interesting recombinant silk-like proteins developed recently include block copolymers with peptide sequences that endow the material with (i) elastomeric properties or (ii) fibronectin cell-attachment sites.[10,11] The latter class of material is expected to have applications in seeding cell growth, say in treatment of burn victims.

Exciting progress is being made with synthetic polypeptides related to mammalian elastins (reviewed in Ref. 13). A pentapeptide sequence repeated in elastin has been shown to be responsible for the elastomeric properties of the natural protein and artificial elastomeric polymers have been produced by cross-linking of filaments formed from chemically synthesized elastin-like peptides.[13-15] Variants of these materials have been shown to contract and relax with changes in temperature and pH. These important discoveries provide a basis for the future development of specialized elastomers differing in thermomechanical and chemomechanical performance properties. Another group has developed a recombinant form of human tropoelastin, the natural precursor to cross-linked elastin.[16] The material was produced in *E. coli* and shown to have biological activity in chemotaxis assays with cultured fibroblast cells. This development opens the way for genetic engineering of novel variants.

Other materials-related proteins under study are human collagen and the water-proof adhesives produced by marine organisms. A synthetic DNA segment encoding a collagen-like tripeptide repeat has been cloned and expressed in *E. coli*, but, disappointingly, this particular product was not stable.[17] The basis of adhesion of the marine protein glues is receiving much attention and is believed to stem from interactions mediated by repetitive amino acid blocks containing one or more lysine or tyrosine residues. Bonding is apparently through interaction of side chain hydroxyl groups formed in tyrosines and (occasional) prolines in the periodic repeats, metal chelation by dihydroxyphenylalanine (DOPA) and cross-linking to lysine amino side groups.[18,19] Early efforts to produce adhesive proteins by recombinant DNA technology have been encouraging.

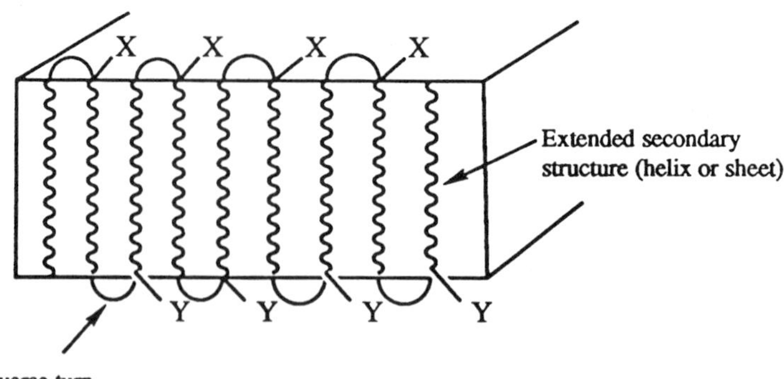

Figure 1. Schematic depiction of a crystalline lamellar material developed from a repetitive polypeptide. The protein features repeating units of β-strands (or α-helices) and linkers that can form reverse turns. Single chains fold to create β-sheets of defined dimension, determined by the length of the chain and resident stems. The anti-parallel β-sheets, in turn, self-associate to form crystalline lamellae. Surface function (X,Y) will be determined by the amino acids in the turn elements.

In one case a mussel cDNA was expressed in yeast to yield an adhesion protein precursor.[20] A second group has expressed a polydecapeptide analog of mussel adhesive protein in *E. coli.*[21]

5.2 *DE NOVO* DESIGN OF PROTEIN MATERIALS

It seems clear now that entirely new classes of protein-based materials can be developed from the first principles and produced by molecular genetic technology. These materials may feature structural or functional motifs that occur in natural proteins, as in some of the examples described above. Alternatively, new polymers may be developed from much less familiar elements. In some cases sequence elements will be selected with the aim of mimicking a domain in a natural protein. In other instances sequences with no biologically distinguishing properties will be selected with a view to eventual use in conditions completely foreign to natural proteins.

De novo development of novel protein materials by genetic synthesis is the thrust of our program. Two classes of protein structures are currently being featured. One corres-

Table-II. Repeating Units of Periodic Polypeptides Currently Under Study

Repeating Unit	**Interest**	**Status**[a]
—(GlyAla)$_x$—GlyProGlu—	Sequence-dependent chain folding	A (x=3, 4, 5, 6,∞)
—(GlyAla)$_3$—XY—	Turn requirement for folded-chain lamellae	A (XY=GlyLeu, GlySer, GlyVal, GlyMet, SerGly, GlyAsp, GlyTyr, GlyAsn, GlyPhe); B (XY = 11 additional pairs)
—(GlyAla)$_x$—GlyGlu—	Correlation of chemical and spatial periodicity	A (x = 3-6)
—(GlyAla)$_3$GlySeMet—[b]	Incorporation of non-natural amino acids	A
—(GlyAla)$_3$GlyGlu(AlaGly)$_3$GlyGlu—	Effect of stem sequence polarity on crystal structure and morphology	A
—(GlyAla)$_3$GlyGlu(GlyAla)$_3$GlyVal— —(GlyAla)$_3$GlyGlu(GlyAla)$_3$GlyMet—	Differentiation of lamellar surface functionality	B B
—Glu$_{17}$Asp—	Monodisperse poly (glutamic acid) and mesogenic poly(γ-benzyl-L-glutamate)	A

[a] A: Protein expressed and isolated;
B: Coding sequence prepared and cloned.
[b] SeMet = L-selenomethionine

ponds to periodic peptides designed to form chain-folded, lamellar structures of defined thickness and surface function.[22-25] The second class consists of helical polymers designed to form liquid crystals.[26] The lamellar materials consist of repeating units of β-strands of variable length joined with amino acid residues predicted to form a reverse β-turn (Figure 1). Single polypeptide chains are expected to form β-sheets which, in turn, are anticipated to form crystalline lamellae through intersheet packing. The liquid crystal materials consist of α-helical protein sequences that currently feature repeating units of polycarboxylic acids.

We have been able to demonstrate the feasibility of our approach through the successful synthesis of a number of materials. The proteins under study and the state of our progress with each are summarized in Table-II. The materials include several lamellar-type proteins and one helix-type polymer. Several of the alanylglycine-rich proteins have been shown to form highly ordered material by X-ray scattering and infrared and NMR spectroscopy, with properties expected of a crystalline β-sheets.[27] Other advances include production of periodic proteins containing the unnatural amino acid seleno-methionine - at sites predicted to be on the lamellar surface[28] and successful expression of a helical polymer of sequence GluAsp$(Glu_{17}Asp)_4$GluGlu.[29] Ongoing efforts are directed at defining effects of stem and turn sequences on chain folding assessing the potential to control hierarchical structure, and attempts to incorporate additional unnatural amino acids into proteins made by recombinant organisms.

6.0 ACKNOWLEDGEMENTS

We are glad to acknowledge the excellent work of several graduate students, postdoctoral fellows and senior guest scientists, especially E.D.T. Atkins, Eric Cantor, Howard Creel, Yoshikuni Deguchi, Michael Dougherty, Michio Kawai, Seenu Kothakota, Mark Krejchi, Kevin McGrath, Ajay Parkhe, and Guanghui Zhang. This work was supported by a grant from the National Science Foundation (DMR 8914359).

7.0 REFERENCES

1. A. J. Anderson and E. A. Dawes, "Occurrence, Metabolism, Metabolic Role, and Industrial Uses of Bacterial Polyhydroxyalkanoates," *Microbiol. Rev.*, **54** [4] 450-472 (1990).

2. H. Brandl, R. A. Gross, R. W. Lenz, and R. C. Fuller, "Plastics from Bacteria and for Bacteria: Poly (β-Hydroxyalkanoates) as Natural, Biocompatible and Biodegradable Polyesters," *Adv. Biochem. Engin/Biotech.*, **41,** 78-93 (1990).

3. A. Steinbuchel, "Polyhdroxyalkanoic Acids" in *Biomaterials*, D. Byrom (ed.) (Stockton Press, New York, 1991) pp. 123-213.

4. *Biomineralization: Cell Biology and Mineral Deposition*, K. Simkiss and K. M. Wilbur (Academic Press, New York, 1989).

5. D. A. Tirrell, M. J. Fournier, and T. L. Mason, "Protein Engineering for Materials Applications," *Curr. Opin. Struct. Biol.*, **1**, 638-641 (1991).

6. M. Xu and R. V. Lewis, "Structure of a Protein Superfiber: Spider Dragline Silk," *Proc. Natl. Acad. Sci. USA*, **87**, 7120-7124 (1990).

7. R. V. Lewis, "Spider Silk: the Unraveling of a Mystery," *Acc. Chem. Res.*, **25**, 392-398 (1992).

8. S. J. Lombardi and D. L. Kaplan, "Isolation, Cloning and Physiological Characterization of Spider Silk from the Golden Orb-Weaver, *Nephila clavipes*," *Polymer Preprints*, **31**, 195-196 (1990).

9. D. L. Kaplan, S. J. Lombardi, W. S. Muller and S. A. Fossey, *"Silks" in Biomaterials*, D. Byrom (ed.) (Stocktron Press, New York, 1991) pp. 1-53.

10. J. Cappello, J. Crissman, M. Dorman, M. Mikolajczak, G. Textor, M. Marquet, and F. Ferrari, "Genetic Engineering of Structural Protein Polymers," *Biotechnol. Prog.*, **6**, 198-202 (1990).

11. J. Cappello, J. Crissman, M. Dorman, M. Mikolajczak, G. Textor, M. Marquet, and F. Ferrari, "The Genetic Production of Synthetic Crystalline Protein Polymers," in *Materials Synthesis Utilizing Biological Processes*, P.C. Rieke, P. D. Calvert, and M. Alper (eds.) *Symp. Proc. MRS*, Vol. **174** (1990) pp. 267-276.

12. D. L. Kaplan and R. V. Lewis, personal communication.

13. D. W. Urry, "Free Energy Transduction in Polypeptides and Proteins Based on Inverse Temperature Transitions," *Prog. Biophys. Mol. Biol.*, **57**, 23-57 (1992).

14. D. W. Urry, "Entropic Elastic Processes in Protein Mechanisms. I. Elastic Structure Due to an Inverse Temperature Translation and Elasticity Due to Internal Chain Dynamics," *J. Protein Chem.*, **7**, 1-34 (1988).

15. D. W. Urry, "Entropic Elastic Processes in Protein Mechanisms, II. Simple (Passive) and Coupled (Active) Development of Elastic Forces," J. Protein Chem., **7**, 81-114 (1988).

16. D. W. Urry, B. Haynes, H. Zhang, R. D. Harris, and K. U. Prasad, "Mechanochemical Coupling in *Synthetic Polypeptides by Modulation of an Inverse Temperature Transition*," *Proc. Natl. Acad. Sci. USA*, **85**, 3407-3411 (1988).

17. Z. Indik, W. R. Abrams, U. Kucich, C. W. Gibson, R. P. Mecham, and J. Rosenbloom, "Production of Recombination Human Tropoelastin: Characterization and Demonstration of Immunologic and Chemotactic Activity," *Arch. Biochem. Biophys.*, **280**, 80-86 (1990).

18. I. Goldberg, A. J. Salerno, T. Patterson, and J. I. Williams, "Cloning and Expression of a Collagen-Analog-Encoding Synthetic Gene in *Escherichia coli*," *Gene*, **80**, 305-314 (1989).

19. J. H. Waite, "Natural Thermoset Proteins," *Polymer Preprints*, **31**, 181-182 (1990).

20. R. A. Laursen, J.-J. Ou, X.-T. Shen, and M. J. Connors, "Characterization and Structure of Mussel Adhesive Proteins," in *Materials Synthesis Utilizing Biological Processes*, P.C. Rieke, P. D. Calvert, and M. Alper (eds.) *Symp. Proc. MRS*, Vol. **174**, (1990) pp. 237-242.

21. R. L. Strausberg, D. M. Anderson, D. Filpula, M. Finkelmen, R. Link, R. McCandliss, S. A. Orndorff, S. L. Strausberg, and T. Wei, in: *Adhesion from Renewable Resources*, R. W. Hemingway, A. H. Conners, and S. J. Branham (eds.) (American Chemical Society, Washington, DC, 1989) pp. 453-464.

22. A. J. Salerno and I. Goldberg, "Expression of a Synthetic Mussell Adhesive Protein in *E. coli*," Proc. *Am. Chem. Soc. Div. Polym. Mat. Sci. Eng.*, **66**, 398 (1992).

23. K. P. McGrath, D. A. Tirrell, M. Kawai, T. L. Mason, and M. J. Fournier, "Chemical and Biosynthetic Approaches to the Production of Novel Polypeptide Materials," *Biotechnol. Prog.*, **6**, 188-192 (1990).

24. H. S. Creel, M. J. Fournier, T. L. Mason, and D. A. Tirrell, "Genetically Directed Syntheses of New Polymer Materials: Efficient Expression of a Monodisperse Copolypeptide Containing Fourteen Tandemly Repeated -$(AlaGly)_4$ProGluGly- Elements," *Macromolecules*, **24**, 1213-1214 (1991).

25. K. P. McGrath, M. J. Fournier, T. L. Mason, and D. A. Tirrell, "Genetically-Directed Syntheses of New Polymeric Materials: Expression of Artificial Genes Encoding Proteins with Repeating -$(AlaGly)_3$ProGluGly- Elements," *J. Am. Chem. Soc.*, **114**, 727-733 (1992).

26. M. J. Fournier, H. S. Creel, K. P. McGrath, M. T. Krejchi, E. D. T. Atkins, T. L. Mason, and D. A. Tirrell, "Genetic Synthesis of Periodic Protein Materials," *J. Bioact. Compat. Polymers,* **6**, 326-338 (1991).

27. (i) M. Krejchi, et.al., and (ii) M. Dougherty, et.al., (both in preparation).

28. M. J. Dougherty, S. Kothakota, T. L. Mason, D. A. Tirrell, and M. J. Fournier, "Synthesis of a Genetically Engineered Repetitive Protein Containing Periodic Selenomethionine Residues," *Macromolecules* (in press, 1992).

29. G. Zhang, M. J. Fournier, T. L. Mason, and D. A. Tirrell, "Biosynthesis of Monodisperse Poly (α, L-Glutamic Acid): Model Rod-Like Polymers," *Macromolecules*, **25**, 3601-3603 (1992).

Subject Index

α-helical conformation, 240
α-helical protein sequences, 271
α-helices, 268, 270
β-crystals, 240
β-pleated sheets, 57, 239
β-sheet, 104, 270, 271
β-sheet conformation, 102, 240
β-strands, 268, 270
β-turns, 268
6M guanidine hydrochloride, 254

A

Absorption spectra, 227, 228
Absorption spectrophotometry, 221, 227
Acoustic emission, 186
Actomyosin, 8
Advanced composites, 163
Aggregation, 229
Al alloys, 36
Algae, 39
Alnico, 205
Alumina, 153
Amino acid composition, 240
Amino acid compositions mollusk shells, 60
Amino acid residues, 248
Amino acids, 245, 255
Amorphous $CaCO_3$, 42
Animal bone implants, 97
Annulus fibrosis, 23
Antheraea pernyi, 237
Antibody-antigen coupling, 100
Antler, 121, 137
Apatite, 117
Apparent density, 123
Aquaspirillum magnetotacticum, 40, 203
Aquatic bacteria, 199
Arachidic acid, 224
Aragonite, 48, 56
Aramid fiber, 176
Araneid Uloborid spiders, 238
Araneus diadematus, 238, 240, 241, 258
Araneus diadematus dragline, 248
Armor, 44, 45, 51
Articular cartilage hydraulic shock absorber, 4
Artificial genes, 264
Artificial membrane interfaces, 217
Aspect ratio, 151
Auditory ossicles, 139

B

Bacillus, 102
Bacteria, 39
Bacterial magnetite, 42
Barium titanate, 156
Bats, 193
Bending moduli, 180
Bessbeetle, 167, 178
Bilayer lipid membranes, 218
Bio-magnetic compass, 199
Bioduplication, 37, 75
Biogenic, 45
Biogenic aragonite, 54
Biogenic silicas, 97
Biological epitaxy, 105
Biological mineralization, 158
Biological reinforcements, 147
Biologically-controlled mineralization, 201
Biologically-induced mineralization, 201
Biomacromolecular phase(s), 45
Biomimetic applications, 190
Biomimetic ceramics, 145

Biomimetic composites, 159
Biomimetics, 1, 11, 195
Biomimicking, 37, 75, 212
Biomineralization, 91, 92, 159, 199
Birefringence, 249
Birefringence fiber, 251
Blubber, 10
BN, B_4C, TiC, ZrO_2, Al_2O_3, 53
Bombyx mori, 237, 239, 249, 255
Bone of birds, 120
Bone of mammals, 120
Bone of reptiles, 120
Bones, 2, 39, 97, 117, 121, 145, 147, 151, 263
Brewster angle, 225
Brick and mortar, 46
Bridging, 49
Butterfly wings, 44

C

Ca-CO_3 interactions, 101
Cadmium-arachidate, 227
Calcareous exoskeletons (corals), 97
Calcite, 48, 56, 92
Calcitic $CaCO_3$, 42
Calcium bones, 131
Calcium carbonate, 93, 147
Calcium content, 119
Calcium oxalate, 93
Calcium phosphate, 93
Ca-PO_4 , 101
Carbohydrate, 263
Carbon fiber, 145
Carboxylate headgroups, 106
Cartilage, 2, 4
Cartilage endplates, 27
Catalyze polymerization, 263
Catastrophic failure, 21
CdS, 40, 97, 217, 223
Cellulose, 146
Cemented carbides, 36
Ceramic, 45
Ceramic-based composites, 45
Ceramic/ceramic, 40
Ceramic/polymer, 40
Chemoreceptors, 192
Chitin, 39, 44, 146
Chitin fibers, 165, 176
Clay, 151
Cloning, 264
Coccolith plates, 97
Coccoliths, 92, 93, 95
Cohesiveness of antler, 139
Cohesiveness of bone, 139
Coincidence lattice sites, 66
Collagen, 4, 39, 117, 145, 146, 147, 263
Collagen composite systems, 13
Collagen fibers, 3, 18, 20, 28
Collagen fibril, 14, 15, 18, 21, 22, 29
Collagenous connective tissues, 2, 3, 13
Colloidal semiconductor particles, 217
Coloring effects, 44
Compartmentalization, 70
Compliance, 125
Composite, 13
Compressed stearate monolayer, 111
Conformation of organic macromolecules, 64
Continuous broad-band, 194
Contractile protein, 8
Control factors in biomineralization, 98
Cornea, 32
Crack blunting/branching, 49
Crack bridging, 49
Creep rupture, 127
Creep rupture experiments, 128, 129
Crimp, 17, 19
Crimp angle, 24
Crimped fibers, 24
Crosslinked polymer networks, 258
Crystal engineering, 91, 108
Crystallization, 257
Crystallography of aragonite, 61
Cubo-octahedral, 201, 208
CuS-Cu_2S, 223

Cuticles, 44, 165

D

Damage, 131
Damage accumulation model, 127
Damage indicator, 139
Damage tolerance, 164, 175, 188
Damage tolerant, 37
De Novo design, 270
Decalcification, 57
Delaminating, 190
Dermal gland, 179
DNA, RNA, 267
Dolphins, 7
Domain walls, 204
Domains, 64
Dragline silk, 237, 238, 243, 245, 247, 250, 255
Dual helicoid, 173, 174, 178
Ductile polymeric matrix, 163

E

E. coli, 269
Ear bones, 139
Echinoderms, 42
Echo-location, 193
Elastic properties, 181
Elastic recoil, 241
Elastin, 146, 248, 254, 263
Elastomeric, 16, 241, 269
Elastomeric fiber, 246
Elastomers, 145
Electromagnetic receptors, 193
Electron-hole confinement, 217
Electron microscopy, 20
Electrostatic accumulation, 102
Elemental maps, 200
Elemental X-ray density maps, 209
Elephant's trunk, 8, 9
Elytra, 170
Emiliania huxleyi, 92, 94
Enamel, 153
Endocuticle, 165
Enzyme engineering, 264
Enzyme regulation, 264
Enzyme-substrate reactions, 100
Epicuticle, 165
Exciton diameter, 217
Exocuticle, 165
Exoskeletons, 44
Extensions, 238

F

Failure stress, 127
Fallow Deer, 120
Faraday cage, 222
Fascicular, 21
Fe(II), 103
Fe(III) oxide, 103
Fe_3O_4, 40
Fe_3S_4, 40, 207
Ferric oxyhydroxide, 201
Ferrihydrite, 103
Ferrimagnetic, 208
Ferrimagnetic greigite, 199
Ferrimagnetic magnetite, 199
Ferritin, 39, 97
Ferroxidase center, 105
FeS_2, 211
Fiber, 175
Fiber pull-out, 175
Fiber-reinforced polymeric composites, 163
Fibril bridging, 175
Fibrillar polysaccharides, 39
Fibrils, 21
Fibrous biopolymers, 238
Filler-matrix interface, 151
First generation twins, 63
Fish, 45
Flagellum, 205
Floatation device, 6
Force-elongation behavior for native viscid silk, 256

Force-elongation curves, 242
Fractal dimension, 136
Fractals, 134
Fracture surfaces, 134, 175
Fracture toughness, 49
Framework macromolecules, 46, 59
Frameworks, 108
Furry fibers, 175
Fused silica, 153

G

Gastropods, 56
Gelantinous nucleus pulposus, 23
Gene sequence, 238
Genetic engineering, 237, 264
Genetic synthesis, 265
Genetically engineer proteins, 263, 265
Genetically engineered enzymes, 265
Geological aragonite, 54, 55
Geomagnetic field, 205
Glass fibers, 151
Glue, 2
Glutamates, 105
Glycine, 14
Glycoprotein, 2
Goethite, 156
Gold citrate, 60
Greigite, 207, 208
Greigite pyrite, 199
Group 11A metal sulfates, 93
Growing edge of a junvenile red abalone, 69

H

H-chain, 105
Haliotidae, 48
Haliotis rufescens, 48, 56
Hard composites, 145
Hard stiff tissues, 39
Haversian system, 151
Heat sink, 6
Helical polymer of sequence, 272
Helical polymers, 271
Helical polypeptides, 14
Helicoidal, 44
Helicospiral, 70
Helix, 17
Hierarchical composite structures, 97
Hierarchical manner, 37
Hierarchical organization, 69
Hierarchical structure, 13, 22, 28, 31
Hierarchical structure - intestine, 29
Hierarchical structure - tendon, 21
Hierarchical twinned, 64
Hierarchical twins, 66
Hierarchically complex, 10
Hierarchy, 150
High hardness ceramics, 53
High temperature superconductors, 36
Histidine, 105
Homeostasis, 211
Host-guest complexes, 100
Human heart valve leaflet, 16
Human intervertebral disc, 16
Human periodontal ligament, 16
Hyaluronic acid, 15
Hydrated, 241
Hydrated silk, 245, 249
Hydrolysis, 157
Hydrophilic bonds, 2
Hydroxyapatite, 117, 145, 147, 152
Hydroxyproline, 14

I

Ideal armor, 52
Impact resistance materials, 51
Indentation, 50
Infrared radiation, 193
Inorganic-organic interface, 91, 98, 106
Inplane moduli, 180
Insect cuticles, 39, 163, 166
Insect exoskeleton, 165
In-situ chemistry, 156
In-situ laminates, 52

Intelligent composite system, 33
Intervertebral disc, 23, 26
Intestine, 13, 27, 28, 30
Intrinsic viscosity, 248
Invertebral disc, 13
Ion-beam milled, 59
Ionotropy mechanism, 61
Iron oxides, 93
Iron sulfide, 206, 211
Iron-reducing bacteria, 201

J

Jade (Jadeite), 153
Jellyfish, 3

K

Kevlar, 175, 238
Kevlar-like super fibers, 258

L

L-chain polypeptides, 105
Laminated composites, 51
Langmuir monolayers stearic acid, 106
Langmuir-Blodgett (LB) films, 218, 225
Laws for complex assemblies, 31
Layered structure, 27
Linearly elastic region, 127
Lipid, 263
Lipid bilayer membrane, 155
Lipid polymers, 267
Liquid crystalline matrix, 40
Liquid crystals, 271
Low-angle boundary, 74
Low-velocity impact, 182
Lubricating fluid, 2
Lycra, 238

M

Macromolecular design, 237
Magnetic, 39
Magnetic domains, 204
Magnetite, 156
Magnetosome chain, 204
Magnetosome membrane, 203
Magnetosomes, 40
Magnetosomes magnetotactic bacteria, 199
Magnetotactic bacteria, 201, 206
Magnetotactic prokaryote, 208
Magnetotactic response, 208
Magnetotaxis, 211
Mammals, 45
Mechanical properties, 117, 118, 241
Mechanical properties of bone, 149
Mechanical properties of nacre, 48
Mechanical properties of silk, 239
Mechanism of growth morphogenesis, 75
Mechanisms of biomineralization, 154
Mechanoreceptors, 192
Mesoglea sea anemones, 3
Microaerophilic bacteria, 208
Microbial culture, 238
Microcrack formation, 49
Microcracks, 119
Microdamage, 117, 131
Microemulsions, 41
Microfiber bridging, 179
Microfibril(s), 14, 15, 175, 176
Micro-Newton forces, 242
Mimetic twin domains, 67
Mineral, 147
Mineral-collagen bonding, 139
Mineral content, 119
Mineral deposition, 267
Mineral-mineral bonding, 139
Mineral volume fraction, 124
Molecular biology, 1
Molecular composites, 42
Molecular genetics, 263
Molecular recognition, 98, 100
Mollusk shell proteins, 109
Mollusks, 45
Monolayer, 223

Monolayer headgroup-aqueous
subphase interface, 219
Mucopolysaccharides, 15
Mucus, 2, 3
Multicellular prokaryote (mulberry), 210
Multifunctional, 37
Multifunctionality, 46
Multilayered composites, 97
Multiple tiles, 66
Muscle, 16
Muscular hydrostats, 8
Mytilus edulis linne, 48

N

NaCl, 110
Nacre, 45, 54, 97, 153
Nacre-like composites, 153
Nanoclusters, 229
Nanocomposite materials, 36
Nanocomposites, 42
Nanophase materials, 108
Nanoscale synthesis, 97
Nanostructural design, 39
Natural composite, 163
Nautilus, 56
Nautilus pompilius, 48
Neodymium-iron-boron, 205
Nephila clavipes, 239
NMR spectroscopy, 272
Nociceptors, 193
Non-circular fibers, 183
Non-crystalline silks, 255
Non-linear optical effects, 36
Non-linear optical properties, 156
Non-linear stress-elongation, 243
Non-linear stress-elongation curves, 247
Non-magnetic pyrite, 199, 211
Non-polar amino acids, 258
North-seeking, 205
Nucleation, 106
Nucleation and crystal growth, 98
Nucleation growth, 46
Nucleator macromolecules, 59
Nucleator proteins, 46
Nucleic acid, 267
Nylon fibers, 146
Nylons, 238

O

Odontotaenius disjunctus, 163, 171
Open-hole compression, 190
Open-hole tension, 188
Optical properties, 242
Optical reflectivity, 221
Optical thickness, 224
Orb web, 238
Organic (polymer) chemistry, 91
Organic ligaments, 49
Organic matrix, 57, 74
Organic membrane, 69
Organic template, 61
Organic/organic biological composites, 44
Organizing proteins, 267
Osmotic shock, 211
Otoconia, 92

P

Particle-reinforced polymeric
composites, 44
PBO, 175
Pentapeptide sequence, 269
Peptide chain architecture, 268
Periodic peptides, 271
Permanent magnetic dipole, 199
Permanent magnets, 205
Phospholipid vesicles, 109
Phospholipids, 108
Photoconductivity, 222, 231
Photoelectrical measurements, 231
Photovoltage, 222
Phylogenetic analysis, 213
Piezoelectric, 156
Pigmentation, 44
Pinctada, 48

Pinctada margaritifera, 48
Plate pull-out, 49
Ply orientation, 174
Plywood architecture, 172
Poisson's ratio, 180
Polarized light, 254
Poly-alanine block, 239
Polyester fibers, 146
Polyethyleneterephthalate, 152
Polyhydroxybutyrate, 96
Polyimide, 145
Polymer materials science, 263
Polymer synthesis, 266
Polymer/polymer, 40
Polymerizable surfactants, 219
Polymerized surfactant vesicles, 218
Polymethylmethacrylate, 158
Polypeptide chains, 248
Polypeptide micelles, 99
Polyphosphate granules, 200
Polysaccarides, 39, 46, 57, 60, 146, 267
Polyvinylidene fluoride, 159
Porosities, 131
Porosity, 121
Post-yield stress, 125
Potassium dihydrogenphosphate, 156
Precipitates, 36
Precipitation in polymers, 157
Preformed holes, 186
Primary structures, 100
Procuticle, 165
Projectile, 51
Proline, 14
Properties of nacre, 153
Protein, 263
Protein-based materials, 268
Protein engineering, 265
Proteins, 39, 46, 57, 146
Proteoglycan, 4, 15
Pseudo-hexagonal arrangement, 65
Pseudo-hexagonal lattice, 66, 106
Pseudo-hexagonal template structure, 66
Pseudomembrane, 70
Pseudomonas bacteria, 266
Pseudomonas oleovorens, 266
Pyrite, 211

Q

Quantum crystallites, 217
Quantum-well structures, 37
Quarternary interactions, 268
Quarternary structures, 102

R

Rabbit cornea, 16
Random-coil molecules, 241, 248
Random links, 249
Random network, 244
Rat intestine, 16
Rat tail tendon, 10, 16
Rat tooth enamel, 154
Recombinant organisms, 272
Recombinant products, 264
Recombinant proteins, 267
Red abalone, 47, 48, 56
Red deer antler, 126
Reflectivity, 221
Regulatory factors, 98
Reinforcing minerals, 145
Resilin, 248, 254
Rice hull, 157
Richardson plots, 134
$ROPO_3$-Ca binding, 101
Rupture, 238

S

Samarium-cobalt, 205
Sarus crane tarsometatarsus, 120
Scaling, 33
Scanning confocal microscope, 139
Scanning tunneling microscopy (STM), 221
Sclerotization, 165
Seashells, 39
Sea urchins, 5, 43
Sea-urchin teeth, 39

Secant modulus, 132
Second generation twins, 63
Secondary and teritary structures, 101
Selenomethionine, 272
Self-assembled, 61
Self-assembly, 46
Self-healing, 37
Semiconductor particle, 223
Semiconductors, 157
Semipermeable flexible tube, 27
Sequence of the proteins, 255
Shape formation (morphogenesis), 46, 67
Shear modulus, 245, 254
Sheet molding compound, 151
Shells, 92, 97, 148, 263
Silica, 156
Silicon carbide, 97
Silicon dioxide, 93
Silk, 146
Silk fibroin-like proteins, 61
Silk fibroin proteins, 44
Silk fibroins, 263, 269
Silk moths, 44
Silk polypeptide backbone, 258
Silkworm's cocoon, 255
Site reactivity, 102
Size quantization, 233
Size-quantized particles, 217
Skin, 193
Sliding, 49
Small particles, 39, 40
Smart aerospace vehicle, 190
Smart skins, 194
Smart technologies, 190
Snakes, 193
Soft biomaterials, 1
Soft connective tissues, 13
Soluble charged macromolecules, 61
Solution chemistry, 157
South-seeking, 205
Space filling tiles, 66
Specific flexural strength, 49
Specific properties, 164
Spiders, 237
Spiders' silks, 237
Spine, 5, 39
Spine of a sea urchin, 6, 42
$SrSO_4$, 110
Stearate monolayer, 107
Stereochemical, 100
Stereochemical matching, 105
Stern layer, 107
Stiffness, 117, 167, 243
Strain-birefringence, 241, 250
Strength, 167, 238
Stress-elongation curve, 243
Stress-elongation viscid silk, 252
Stress-strain behavior of rat tail tendon, 22
Stress-strain behavior of the intestine, 30
Stress strain curve for bone, 133
Stress-strain curves, 126
Stroma, 31
Strong polymers, 145
Structural framework, 267
Structural hierarchy, 168
Submucosa, 27
Superalloys, 36
Supermodulus effect, 36
Superparamagnetism, 204
Superplastic, 53
Superstructure, 66, 72
Supramolecular assemblies, 98
Surface pressure-surface area isotherm, 224
Surfactant molecules, 106
Surgical implants, 159
Synthetic biological ceramics, 152
Synthetic biological composites, 149
Synthetic chemistry, 91
Synthetic laminated FRPC, 172
Synthetic polymers, 151

T

Teeth, 39, 97
Template activity, 268
Tendon, 13, 16, 19, 20, 28, 44
Tensile fibers, 237
Tensile modulus, 245
Tensile strength, 117
Tertiary, 268
Thermoelastic, 241
Thermoreceptors, 192
Thermosensitive, 194
Third generation twins, 63
Three-dimensional quantum size effect, 217
Three polypeptide chains, 15
Titania, 156
Tooth, 148
Tortuosity, 49
Toughening mechanisms, 49
Toughness, 117, 130, 238
Transmission electron microscopy (TEM), 221
Transparent composites, 159
Trilobites, 92
Tropocollagen, 14, 15
Tropocollagen helices, 15
Tropocollagen helix, 21
Tympanic bulla, 139
Tympanic bulla of the whale, 127

U

Ultimate strain, 123
Ultimate strengths, 181
Ultimate stress, 125
Ultramicrotomed section, 59
Ultrastructural assembly, 97
Unnotched specimens, 189
Unsymmetrical layup, 178
Uranyl oxide, 109

V

Vaterite, 110
Vertebrate bony hard tissues, 117
Vesicles, 41, 108, 155
Viscid silk, 237, 238, 253, 255

W

Wallaby femur, 137
Water, 117, 147, 241
Water content, 27
Webs of spiders, 44
Whale blubber, 6
Whales, 7
Whitening, 140
Wood, 263
Work of fracture, 117, 130, 136

X

X-radiographs, 189
X-ray density maps, 201
X-ray diffraction, 239
X-ray scattering , 272

Y

Yield strain, 122, 123, 125
Yield stress, 121
Young's modulus, 119, 121

Z

Zero-dimensional excitons, 217
Zinc-arachidate, 227
Zirconia, 156
ZnS, 217, 223